T0181011

Deep-Buried Large Hydrocarbon Fields Onshore China: Formation and Distribution

Suyun Hu · Tongshan Wang

Deep-Buried Large Hydrocarbon Fields Onshore China: Formation and Distribution

Suyun Hu
PetroChina Research Institute of Petroleum
Exploration and Development
Beijing, China

Tongshan Wang
PetroChina Research Institute of Petroleum
Exploration and Development
Beijing, China

ISBN 978-981-16-2287-8   ISBN 978-981-16-2285-4   (eBook)
https://doi.org/10.1007/978-981-16-2285-4

Jointly published with Petroleum Industry Press
The print edition is not for sale in China (Mainland). Customers from China (Mainland) please order the print book from: Petroleum Industry Press.

Translation from the Chinese Simplified language edition: *Zhongguo lushang shenceng dayouqitian xingcheng yu fenbu* by Suyun Hu, and Tongshan Wang, © Petroleum Industry Press 2019. Published by Petroleum Industry Press. All Rights Reserved.

This Springer imprint is published by the registered company Springer Nature Singapore Pte Ltd.
The registered company address is: 152 Beach Road, #21-01/04 Gateway East, Singapore 189721, Singapore

Preface

With the development of hydrocarbon exploration, it is difficult to make a major breakthrough in the middle-shallow strata. As a result, the hydrocarbon exploration in deep strata is a great significance to increase reserves and stabilize productvity in the hydrocarbon fields. In recent years, the deep hydrocarbon exploration has become a global trend, and the discovery of high-temperature ultra-deep hydrocarbon reservoirs challenges the traditional understanding of hydrocarbon geology. It is a great academic significance and economic value to carry out the deep hydrocarbon geological research.

This book provides theoretical directions and technical supports for the breakthrough of deep hydrocarbon exploration and the increase in the reserves of scale by systematically analyzing and summarizing the formation conditions, distribution rules and main controlling factors of deep hydrocarbon fields, and predicting the exploration prospect of deep hydrocarbon onshore in China based on the exploration research and latest progress of deep hydrocarbon in key petroliferous basins onshore such as Sichuan Basin, Tarim Basin and Ordos Basin in recent years. This book focuses on the analysis and discussion of the hydrocarbon generation mechanism in deep source rocks, the rules of hydrocarbon accumulation in ancient strata during the tectonic period, the hydrocarbon accumulation potential in gypsum-carbonite symbiosis and subsalt strata, the main controlling factors and distribution rules of deep hydrocarbon fields, and the significant superseding layers in deep strata. The geological understandings of the formation and distribution of deep large hydrocarbon fields are preliminarily obtained, that is, full hydrocarbon supplied by two types of source kitchens, large-scale reservoirs formed by three types of rocks, hydrocarbon accumulation controlled by three types of ancient source rocks and hydrocarbon accumulation crossing major tectonic stages.

This book is divided into five chapters: Chap. 1 is written by Hu Suyun, Wang Tongshan, Xu Zhaohui, Zhao Xia, etc.; Chap. 2 is written by Wang Tongshan, Li Yongxin, Qin Shengfei, Chen Yanyan, Ma Kui, Fang Jie, etc.; Chap. 3 is written by Liu Wei, Jiang Qingchun, Shi Shuyuan, Li Qiufen, Huang Qingyu, Wang Kun, Zhao Zhenyu, He Youbin, Zhang Yueqiao, Zhai Xiufen, Bai Bin, etc.; Chap. 4 is written by Wang Tongshan, Liu Wei, Jiang Hua, Bo Dongmei, He Dengfa, etc.; and Chap. 5 is written by Wang Zecheng, Xu An Na, Jiang Qingchun, Gu Zhidong, Shi Shuyuan,

Xu Zhaohui, Lu Weihua, Li Jun, Yuan Miao, Fu Ling, Lin Tong, Sun Qisen, etc. This book is finalized by Hu Suyun and Wang Tongshan.

Academician Zhao Wenzhi, Prof. Gao Ruiqi, Prof. Gu Jiayu, Prof. Luo Ping, Prof. Zhang Baomin and other experts put forward valuable suggestions for the preparation and examination of the manuscript, and we express our heartfelt thanks here.

Because of the complexity of deep hydrocarbon exploration research and the limited level of writers, there are many inadequacies in the book, so criticism from readers are warmly expected.

Beijing, China   Suyun Hu
                                                                       Tongshan Wang

About This Book

The prospect of onshore deep oil and gas exploration in China is predicted in this book by the analysis on major formation conditions of deep oil and gas reservoirs, major control factors and distribution rules of deep large oil and gas fields, based on the two major scientific issues in key onshore petroliferous basins of China, such as the hydrocarbon generation and accumulation of deep oil and gas.

This book can be used as a reference or auxiliary teaching material for many majors in colleges and universities, such as geological resources and geological engineering, resource exploration engineering, petroleum engineering and others as well as for scientific researchers and production technicians engaged in oil and gas exploration and development.

Cataloguing in Publication (CIP)
Formation and distribution of deep large oil and gas fields onshore in China /Hu Suyun, et al.—Beijing: Petroleum Industry Press. 2019.7 ISBN 978-7-5183-3432-2 I. ① China…II. ① Hu…III. ① Onshore oil and gas fields—formation of oil and gas reservoirs—research—China ② Onshore oil and gas fields—oil and gas reservoirs—distribution law—research—China IV .① P618.130.2

Archival Library of Chinese Publications CIP data kernel word (2018) number 101696
Publication and Distribution: Petroleum Industry Press
(100011, No. 1, Anhuali 2 district, Anding gate, Beijing)
Website: www.petropub.com
Editorial department: (010) 64523543 Book Marketing Center: (010) 64523633
Distribution: National Xinhua Bookstore
Printing: Beijing CNPC Color Printing Co., Ltd.
First edition in July 2019. First priting in July 2019
787 × 1092 mm
Book size: 1/16
Printed sheet: 12.5
Words: 240 thousand words
Price: RMB100.00 yuan

(In case of printing and packaging quality problems, the book marketing center of Petroleum Industry Press shall be responsible for the replacement.)

Contents

Chapter 1
Introduction

With the increasing pressure of oil and gas supply and energy security risks, it has become an inevitable trend for oil and gas exploration to deep and ultra-deep strata. In recent years, with the improvement of technology in exploration and development, oil and gas reservoirs distributed in extreme environments such as deep sea, deep strata and polar regions have been identified as the main directions of exploration and development, and a number of large-middle oil and gas fields have been discovered in the deep strata. Compared with the middle-shallow oil and gas accumulation theory, the deep oil and gas accumulation theory is not mature enough. Therefore, in order to meet the basic status of domestic oil and gas resources supply, we must promote the development of deep oil and gas exploration through the systematic summarization and research.

1.1   Exploration and Geological Characteristics of Global Deep Hydrocarbon Fields

Despite the late beginning of exploration in deep hydrocarbon fields, the great discoveries and breakthroughs have been made in petroliferous basins all over the world, and the oil and gas production is increasing year by year. The formation conditions of deep oil and gas fields are different from those in the middle-shallow strata. In the future, with the continuous process in geological theories and the exploration and the development of technology in deep oil and gas, the deep oil and gas reserves will be substantially improved.

© Petroleum Industry Press 2021
S. Hu and T. Wang, *Deep-Buried Large Hydrocarbon Fields Onshore China:
Formation and Distribution*, https://doi.org/10.1007/978-981-16-2285-4_1

1.1.1 Definition of Deep Strata

As for the definition of deep strata, there is no strict international standard and different countries and institutions have different views (Hansheng and Feng 2000; Ministry of Land and Resources 2005; Wenzhi et al. 2014; Guangya et al. 2015). Russia defines deep strata as exploration depth greater than 4,000 m, the United States and Brazil define deep strata as exploration depth greater than 4,500 m, and Total Company defines deep strata as exploration depth greater than 5,000 m. In 2005, "Regulation of Petroleum Reserves Estimation" was issued by National Commission of Mineral Reserves, which defined deep strata as a depth of 3,500–4,500 m, and ultra-deep strata as a depth of more than 4,500 m. China Drilling Project defined deep strata as a depth of 4,500–6,000 m, and ultra-deep strata as a depth of more than 6,000 m. Based on the variation of temperature and pressure and exploration practice in the eastern and western regions of China, this book defined the strata with the buried depth of 3,500–4,500 m in the eastern area as the deep strata, and over 4,500 m as ultra-deep strata; the strata with the buried depth of 4,500–6,000 m in the western area as the deep strata, and the strata with buried depth over 6,000 m as ultra-deep strata. Deep strata have both the concepts of depth and stratigraphy, "deep" and "ancient" are the basic characteristics. According to this definition, most of the important discoveries of domestic oil and gas exploration in recent years belong to the fields of deep to ultra-deep strata.

1.1.2 Distribution and Exploration of Global Deep Hydrocarbon Fields

(1)   History and situation of deep oil and gas exploration

In recent years, although a number of large and middle oil and gas fields have been discovered in the world, deep oil and gas exploration can be traced back to the 1950s. In 1956, the first deep gas reservoir in the world was discovered in the middle Ordovician carbonate rocks at a depth of 4,663 m, the Carter Knox Gasfield of Anadarko Basin, USA. In 1977, with the breakthrough of drilling and completion technology in deep strata, a dolomite gas reservoir was found in Arbuckle Group of Cambrian and Ordovician strata, Mills Ranch Gasfield, at a depth of 8097 m. In 1984, an oil reservoir was discovered in Permian dolomite reservoir of Villifortuna Trecate Oilfield, Italy, at a depth of 6,400 m. Since 1980, deep oil and gas exploration has gradually expanded from land to sea, such as the discovery of gas reservoirs in limestone reservoir of Permian Khuff Formation, Fateh Gasfield, Arabian basin, at a depth of 4,500 m in 1980. At present, the global deepest oilfield is Anchor Oilfield, USA, with a depth of more than 10,000 m and the proved reserves of 44×10^8 t; the deepest gasfield in the world is Mills Ranch Field, USA, with a depth of 7,663–8,083 m and the proved reserves of 112×10^8 m^3. Important breakthroughs of deep oil and gas exploration have been made in the deep and ultra-deep water regions, such as Mexico Gulf, eastern Brazil and West Africa, as well as the Arctic region (Timan-Burchella Basin in Russia) (O'Brien and Lerche 1988; Lerche and Lowrie 1992; Aase and Walderhaug 2005; Ehrenberg et al. 2008; Ajdukiewicz et al. 2010; Guangming et al. 2012; Cao et al. 2013; Guoping and Binfeng 2014).

The global production of deep and ultra-deep oil and gas have been increased year by year. At present, the proportion of proven reserves and production of deep oil and gas is relatively small but growing faster (Fig. 1.1). In recent years, the number of oil and gas reservoirs discovered in ultra-deep strata with more than 6,000 m in the world has been increased significantly. Since 2000, 106 oil and gas fields have been

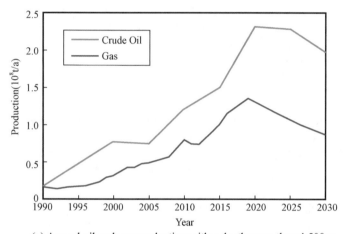

(a) Annual oil and gas production with a depth more than 4,500 m

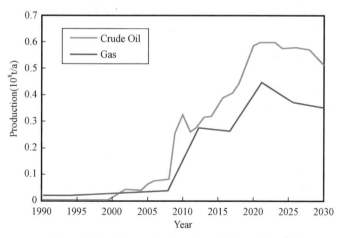

(b) Annual oil and gas production with a depth more than 6,000 m

Fig. 1.1   Prediction of global deep oil and gas production (according to Xiaoguang et al. 2014)

discovered in the ultra-deep strata buried more than 6,000 m in marine area; 51 oil and gas fields have been discovered in the ultra- deep strata buried more than 6,000 m in land area. By the end of 2015, 178 industrial oil and gas fields with buried depth greater than 6,000 m have been found in the world (according to IHS). According to the preliminary results of resource evaluation, the resource extent of deep oil and gas undetected is 345×10^8 t with huge exploration potential. According to IHS statistics in 2013, the deep proved reserves of crude oil is 115.5×10^8 t and natural gas is 76×10^8 t (oil equivalent), accounting for 3.3% and 3.2% of the global total oil and gas reserves respectively. Among them, the ultra-deep proven oil reserves of crude oil is 15×10^8 t and of the natural gas is 6.2×10^8 t (oil equivalent), accounting for 13% and 8.2% of the total deep oil and gas reserves respectively. According to Wood Mackenzie's statistics in 2013 (Guangya et al. 2015), in 2012, the global annual production of crude oil in reservoirs deeper than 4,500 m is 1.1×10^8 t, accounting for 2.7% of the global total crude oil production, mainly in the Gulf of Mexico, Kazakhstan and Brazil. The oil production in the depth of more than 6,000 m is 0.26×10^8 t per year, accounting for 0.63% of the global total production, mainly from the Gulf of Mexico in the United States. In 2012, the global production of natural gas from reservoirs deeper than 4,500 m is 0.77×10^8 t (oil equivalent), accounting for 2.86% of global production, mainly from the D6 area of KG basin in India, followed by the Gulf of Mexico in the United States. The natural gas from reservoirs deeper than 6,000 m is 0.25×10^8 t (oil equivalent), accounting for 1% of global production, mainly from the D6 area of India and the Shah Deniz Gasfield of Azerbaijan.

(2)   Distribution of deep oil and gas resources

The global deep oil and gas resources are mainly distributed in the Gulf Coast Basin, Permian Basin, Anadarko Basin, Rocky Mountain Basin, California Basin and Alaska Basin in the North America; Maracaibo Basin, Santa Cruz-Tariji Basin and Sureste Basin in the Central and South America; Dnieper-Donets Basin, Vilyuy Basin, North Caspian Basin, South Caspian Basin, Middle Caspian Basin, Amu-Darya Basin, Azov-Kuban Basin and Fergana Valley Basin in the former Soviet Union; Po Vally Basin and Aquitainethe Basin in Europe; Oman Basin in the Middle East and the Sirte Basin in Africa (Wang Yu et al. 2012).

   According to the statistical data (Guangya et al. 2015), among the deep oil and gas resources discovered in the world, the proved deep oil and gas reserves in Latin America are the largest, accounting for 65% and 37% of the total deep oil and gas reserves respectively (Fig. 1.2). The proved and controlled recoverable reserves in North America are 38.28×10^8 t (oil equivalent), most of which are distributed in the deep and ultra-deep water areas of Mexico Gulf. The region with the largest amount of deep natural gas and condensate discovered is in the Middle East, which is about 34.20×10^8 t (oil equivalent), and 56% of deep natural gas is in Arabian Basin. In terms of depth and strata (Fig. 1.3), the discovered reserves decrease with depth. The deepest oil and gas fields are more than 10,000 m, and the oil and gas reserves discovered between 4,500 and 5,500 m are accounting for 80% and 84%

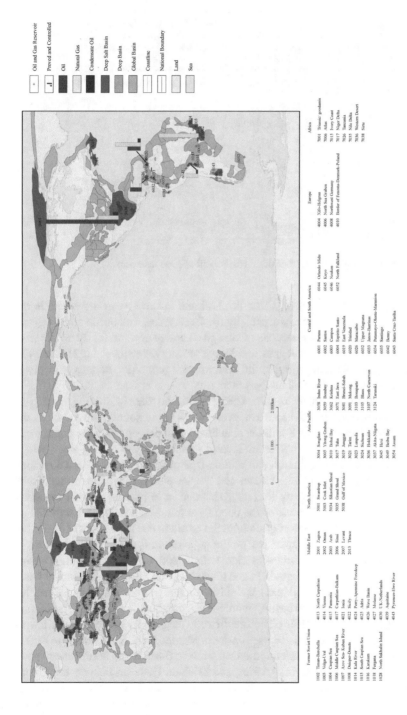

Fig. 1.2 Global distribution of deep petroliferous basins and deep reservoirs (Guoping and Binfeng 2014)

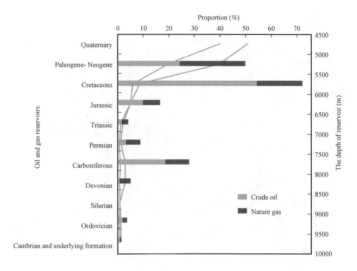

Fig. 1.3 Distribution depth and strata of global deep oil and gas reservoirs

of the total reserves respectively. The deep oil and gas reserves concentrate in the Mesozoic and Cenozoic strata, especially the Cretaceous, Paleogene and Neogene strata, in which the deep Cretaceous oil and gas reserves account for 48 and 24% of the total reserves, and the Pleogene and Neogene reserves account for 21 and 34%. At the same time, with the aging of the reservoir, the proportion of deep natural gas in the total amount of deep oil and gas increases.

Generally speaking, there are some characteristics in these basins: (1) the deposit thickness is generally more than 8 km, even 20–25 km in some basins, such as the North Caspian Basin and the South Caspian Basin with the deposit thickness of more than 25 km; (2) most of the basins are located in ancient hydrocarbon areas, which also have developed a large number of middle-shallow hydrocarbon reservoirs, such as Gulf Coast Basin in the United States and Pre-Caspian Basin in the former Soviet Union, which are famous petroliferous basins and have been developed for many years (Yu et al. 2012); (3) most of the deep reservoirs are developed below 5500–6000 m. At present, the deepest gasfield in the world is the Mills Ranch Gasfield in Anadarko Basin, and the deepest gas producing stratum is the dolomite stratum in Arbuckle Formation of Cambrian-Silurian, with a depth of 7965 m. (4) these basins usually have low geothermal gradient and high abnormal pressure. In terms of basin types, the discovered deep oil and gas fields are mainly concentrated in the passive continental margins, foreland basins, middle-lower combinations of craton basins and rift basins (Table 1.1). The deep oil and gas fields in the global passive marginal basins are mainly found onshore and deep-sea basins of Mexico Gulf, Santos Basin in Brazil, Levitan Basin in the eastern Mediterranean, Browse Basin in the northwestern shelf of Australia, Niger Delta Basin in Africa, and Krishida-Gedavari Basin in the eastern sea of India. These deep oil and gas fields are mainly distributed in deep and ultra-deep water areas and composed of sandstone reservoirs in Mesozoic and Cenozoic.

Table. 1.1 Distribution of major oil and gas exploration fields in deep strata of the world

Exploration field	Basin		Reservoir	Buried depth of reservoir (m)	Typical oil and gas fields
Middle-lower Assemblages of Craton Area	Cratonic Margin	Arabian Platform Basin	Sandstone of Devonian Juaf Formation	4900	Ghawar Oilfield
			Jurassic Carbonate Rocks	4570–4920	Umm Niqa Gasfield
		Amu Darya Basin	Jurassic Reef Limestone	4295–4795	Yashlar Gasfield
	Intracatoinc Basin	Pre-Caspian Basin	Carboniferous Under-salt Limestone	3900–4600	Casagan Oilfield
		Permian Basin	Silurian Carbonate Rocks	4785–5715	Vermejo/moore-Hopper Gasfield
Foreland Basin	Zagros Fold Belt		Mesozoic Carbonate Rocks	4500–5200	Ramin Oilfield, Marun Gasfield
	Anadarko		Silurian-Devonian Carbonate Rocks	5395–6001	Mills Rance Gasfield
	East Venezuela		Cretaceous Sandstone	4650	Santa Barbara Oilfield
	Chaco		Devonian Sandstone	4410–5037	San Alberta Gasfield
	Deep South Caspian Sea		Pliocene Sandstone	5600–6265	Shah Deniz Gasfield
Passive Marginal Basin	Drift Sequence	Surest, Mexico	Jurassic-Cretaceous Carbonate Rocks	3900–4720	Bermudez Oilfield
		Deep sea in the Gulf of Mexico	Turbidite Sandstone in Paleogene Deep Sea	>8000	Tiber Oilfield
		Krishida-Gedavari	Cretaceous Turbidite Sandstone	4205–5061	Deen Dayal Gasfield

(continued)

Table. 1.1 (continued)

Exploration field	Basin		Reservoir	Buried depth of reservoir (m)	Typical oil and gas fields
	Rift Sequence	Blaus	Middle Jurassic Sandstone	5012	Poseidon 1 Oilfield
		Santos, Brazil	Cretaceous Shell Limestone	4900	Lula Oilfield
Rift Basin	Central Graben of North Sea Basin		Jurassic Sandstone	5350–5630	Elgin-Franklin Gasfield
	Bedrock in Jiulong Basin		Bedrock Fracture in Mesozoic	4500–5500	Bach Ho Oilfield

Deep craton oil and gas fields are mainly composed of reef carbonate reservoirs and stratigraphy-structure traps, which are mainly distributed in Arabian Platform Basin in Middle East, the under-salt strata of Pre-Caspian Basin, the under-salt strata of Amu Darya Basin and Permian Basin in the United States. Deep oil and gas fields in foreland basins mainly develope carbonate rocks and sandstone reservoirs, which are dominated by structural traps related to folds and thrust faults, such as the Zagros fold belt, Malacaibo Basin in Andean Foreland, the east Venezuela Basin, Chaco Basin, South Caspian basin, Anadarko Basin, etc. The deep oil and gas fields in the rift basins are mainly distributed in the central North Sea Graben and Vienna Basin, and the deep sandstone reservoirs are characterized by high temperature and high pressure.

1.1.3 Formation Conditions of Global Deep Oil and Gas Fields

The hyrdrcarbon geology theories and understandings developed in hydrocarbon exploration of middle-shallow strata are limited in deep hydrocarbon exploration. According to the classical petroleum geology theory (Tissot and Welte 1978; Wenzhi et al. 2005), most of the discovered oil in the world exists within the "liquid window" (the temperature range from 65.5 to 149 °C), above which oil will be replaced by natural gas. However, deep oil and gas exploration has confirmed that the temperature at which oil exists is well above the limit (Jincai et al. 1999). In some reservoirs, liquid hydrocarbon accumulations can still exist in the oil fields at 295 °C, such as the North Sea Oilfield, the Washington Oilfield in America, the Barr Lake Oilfield, the Paladin Oilfield, the Leyik Oilfield and Bieer Oilfield in Mexico Gulf Basin, the Marun Oilfield in the Persian Gulf, and even the Bla Sea reservoir in the Pre-Caspian Basin in Russia. The experiences and theories summarized in the exploration in deep oil and gas fields by making a rapid development of theoretical research and a full understanding of generation, migration, preservation and distribution in deep oil and gas reservoirs can lead to a better exploration of oil and gas.

So far, a great deal of researches on deep oil and gas exploration have been made by many scholars (Wenzhi et al. 2007; Wenhui et al. 2009; Pang Xiongqi et al. 2010). These researches mainly have focused on the following aspects: the temperature of oil and gas in deep strata, the material base of deep oil and gas reservoir formation, the stability of deep oil and gas, main controlling factors of deep reservoir property, the influence on oil and gas accumulation caused by abnormal high pressure and the analysis of geological conditions in typical deep oil and gas fields. Through the analysis of the reservoirs geological factors in many global deep basins, the following conditions are required for the formation of deep oil and gas fields.

(1)   High-quality source rocks

Like the oil and gas in middle-shallow strata, as the material basis, the source rocks with organic matters are also needed in deep oil and gas reservoirs for hydrocarbon generation. The existence of deep source rocks is an indispensable condition for the formation of deep reservoirs. The ultra-deep source rocks with high content of organic carbon in the petroliferous basins distributed widely and the it is mainly composed of terrigenous clastic rocks and carbonate rocks with the organic carbon content from 0.25 to 6%. The organic carbon content of the ultra-deep source rocks is mainly controlled by the sedimentary facies and the organic matter and has no relation with the burial depth. In addition to temperature and pressure, the maturity of the ultra-deep source rocks is related to the rate of the basin. Compared to the basin with essentially constant subsidence, the mature period of the source rocks with late-rapid subsidence is later, and the rate of hydrocarbon generation is higher.

(2)   Good reservoir-cap combination

Many types of reservoirs, such as the pore type, fracture type, karst cave-fracture type, pore-fracture type and other types, exist in deep reservoirs of clastic rocks and carbonate rocks. Compared to the reservoir in shallow-middle depth, the porosity of the ultra-deep reservoir is not relatively low, land mainly composed of secondary pore. The reservoir property of the ultra-deep reservoir is not only controlled by the pressure and temperature, but also the stress. Under the overpressure environment, the compaction, cementation and dissolution are reduced, so the reservoirs with relatively high porosity and permeability are developed in the ultra-deep reservoirs. In the ultra-deep oil and gas reservoirs, the proportion of gas reservoirs and condensate gas reservoirs is obviously increased. High-quality reservoirs can be developed in the deep strata and filled by oil and gas under the conditions of favorable sedimentary facies, supergene weathering and leaching, dissolution of cement and dolomitization in diagenesis, abnormal high pressure, early oil and gas injection and fracture development.

   In deep reservoirs, the existence of good reservoir-cap combination, especially regional caprock, is the key controlling factor of preservation as the hydrocarbons are easily lost upward or cracked into gas due to the deep depth and high temperature. The high-quality caprocks of ultra-deep reservoirs are mainly salt rock and mudstone. With the characteristics of compactness, deformability and strony toughness, the salt rocks are the best cap rocks for ultra-deep reservoirs, especially for large-scale reservoirs, such as Tengiz Oilfield in Pre-Caspian Basin (Huangjuan et al. 2016). The reservoir scale may increase with the depth if there are high-quality caprocks in the basin.

(3)   Abnormal pressure

The statistics of global deep oil and gas show that the abnormal high-pressure are common developed in the areas where deep oil and gas accumulated, and overpressure controls both the thermal evolution of deep source rocks and the porosity and permeability of reservoirs (Hunt 1990; Caillet et al. 1997; Wilkinson et al. 1997; Haiqing et al. 1998; Fang et al. 2002; Hao et al. 2007; Jiarui et al. 2016).The deep source rocks entered the stage of quasi-metamorphism in the traditional model may be still in the favorable stage of hydrocarbon generation and expulsion, and become the effective source rocks for hydrocarbon accumulation in deep strata, as the increased abnormal pressure can restrain obviously the thermal evolution of organic matter and the generation of oil and gas. The development of overpressure can maintain the porosity and permeability of the deep reservoir and provide reservoir conditions for deep oil and gas accumulation, because the effective stress of the overpressure system is reduced, which result in the weakening of compaction and suppression of pressure dissolution, and make the deep reservoir with higher porosity and permeability.

(4)   Favorable tectonic setting

Special oil and gas reservoirs can be formed on the special tectonic setting as the source rocks and reservoir-cap combination of deep oil and gas are controlled by the development and type of the basin, as well as the filling of the sedimentary facies belts (Jianghai et al. 2014). Passive continental margin, foreland basin, middle-lower assemblage of craton basin and rift basin are favorable basins for the development of deep oil and gas fields, because: (1) a thick sedimentary layer of the material conditions for deep hydrocarbon generation and preservation can be formed; (2) abnormal high pressure can be easily formed, which restrains the generation and expulsion of hydrocarbons and makes the depth of oil generation window decreases, meanwhile, the overpressure in the reservoir makes the reservoir maintain better porosity and permeability conditions; (3) these types of basins are prone to form a large number of fractures, which increase the porosity and permeability of the reservoirs and promote the expulsion and accumulation of oil and gas; (4) a large number of structural traps in rift basins and foreland basins are formed, especially those related to faults and anticlines which can form good trap conditions.

Nowadays, the proportion of deep oil and gas fields in total reserves is increasing. In the United States, the average hydrocarbon reserves in deep strata have exceeded that in middle-shallow strata. In Russia, the average hydrocarbon reserves in deep strata are equal to that in middle-shallow strata (Shixin et al. 2005). It can be predicted that the total hydrocarbon reserves of deep strata will be greatly increased with the technology development of deep hydrocarbon exploration.

1.2   Exploration and Challenges of Onshore Deep Oil and Gas Fields in China

Deep oil and gas exploration in China is just beginning. In recent years, with the continuous effort on oil and gas exploration in deep and ultra-deep strata, a series of large-scale oil and gas fields have been discovered and it shows great exploration potential. At the same time, a series of new understandings and developments have

been made in the generation and preservation of deep oil and gas, reservoir formation mechanism, evaluation and exploration potential of hydrocarbon resources, and
exploration engineering technology. The research of deep oil and gas geology is
incomplete yet, and the theories that can be directly responsible for the guidance of
hydrocarbon exploration and production have not been formed. Lots of key geological problems and evaluation technologies are still under exploration and need to be
solved urgently.

1.2.1 Exploration History and Current Situation of Deep Oil and Gas Fields in China

The exploration of deep oil and gas reservoirs in China began in the 1970s and
1980s. In 1966, the first deep well Songji 6 (4719 m) was drilled in China, and the
first ultra-deep well Nvji (6011 m) was drilled in 1976, and the well Guanji (7175 m)
was drilled in 1978. The important information of deep strata in the valuable deep
wells laid a solid foundation for the geology of deep oil and gas in China.

In the late 1980s, the research on deep strata were carried out by petroleum
geologists since the development of eastern major oilfields gradually entered the
middle-late stage (Shixin et al. 1999; Zhiyi 2005; Zhongjian and Hui 2009). During
this period, after several exploration meetings in the Northeast of China and Tarim
Basin, breakthroughs were made in well Shacan 2 and well Kela 2, and high-quality
unitized oilfields such as Lunnan, Tazhong, Donghetang and Hudson were discovered. And the theories of deep oil and gas have been developed rapidly since the
exploration potential of deep oil and gas has been valued by explorationists.

In the twenty-first century, with the continuous increase of national energy demand
and the depletion of oil and gas production in middle-shallow strata, China and many
oil companies have increased their support for deep oil and gas geological research
and exploration and production, as deep oil and gas play an important role in the
development of hydrocarbon industry and the increase in reserve and production.
Mang "973" projects and national key hydrocarbon projects have been set up to
promote the development of deep hydrocarbon geology. A series of theories which
are important for deep oil and gas exploration are put forward by researchers in
hydrocarbon geology, such as the "successive gas generation" theory of organic
matters in the study of source rocks, "bimodal pattern" theoretical model of hydrocarbon generation for high-overmature source rocks, the theory of multiple hydrocarbon generation for marine source rocks (Zecheng et al. 2002; Wenzhi et al. 2005;
Jinxing et al. 2008; Zhaoyun et al. 2009; Wenhui et al. 2009, 2012). In the reservoir
research, it is proposed that the bedding karstification and interlayer karstification
are the important generation mechanism of large-scale effective reservoirs in the
ancient carbonate rocks. Under special geological conditions, the deep clastic rocks
develope abnormal high-porosity interval and secondary-porosity zone, and volcanic

rocks develope two types of effective reservoirs, primary reservoir and secondary-weathering reservoir. In the research of hydrocarbon accumulation, it is proposed that the liquid window can be maintained for a long time through progressive burial and the coupling of annealing and heating, meanwhile several "golden zones" of exploration can be developed in deep superimposed basins with multiple hydrocarbon source kitchen, multistage reservoirs, multistage accumulation and late availability. The hydrocarbon reservoirs of deep strata in superimposed basin and marine strata greatly enrich the theory of deep oil and gas geology, with the characteristics of "hydrocarbon controlled by source rocks and cap rocks, and hydrocarbon accumulation controlled by slope". At the same time, a number of academic works focused on deep oil and gas exploration have been published one after another, including representative works such as "Deep gas fields in China", "Deep petroleum geology in eastern China", "Deep fluid activities and effects of oil and gas accumulation" (Jin Hansheng 2002; Tao 2002; Zhijun et al. 2007; Xiaorong et al. 2016). These monographs have studied the geological conditions, the hydrocarbon generation and evolution, the accumulation model and distribution of middle-large deep oil and gas fields in China from different perspectives.

In recent years, a series of significant breakthroughs and discoveries gas have been made in deep oil and gas exploration of key petroliferous basins in China. Marine carbonate oil and gas fields such as Lunnan, Tahe and Tazhong, and continental clastic gas fields such as Dabei and Keshen are found in the Tarim Basin (Zhijun 2005; Yuzhu 2008; Zhongjian and Hui 2009; Guangyou et al. 2010; Zhengzhang et al. 2011; Wenzhi et al. 2012; Chunchun et al. 2017). Among which, Donghetang Oilfield is the deepest one, with a maximum reservoir depth of 6130 m, and the proved reserves of 3251×10^4 t. In addition, the reservoirs with buried depth more than 5000 m exist in Yaha Oilfield, Sangtamu Oilfield, Yangtake Oilfield, Tahe Oilfield, and Yangtake Gasfield. The maximum buried depth of Carboniferous volcanic reservoir in Shixi Oilfield, Junggar Basin is 4530 m (Jinghong et al. 2011). The deep reservoirs with buried depth of 3500–4000 m are continuously found in Karamay Oilfield, Chepaizi Oilfield and Mabei Oilfield of northwestern Junggar Basin, Hutubi Gasfield, Kayindike Oilfield in the south of Junggar Basin, Mosuowan Uplift and Mobei Uplift in central Junggar Basin. Shunbei Oilfield, which was discovered in 2016, has a resource of 17×10^8 t, including oil of 12×10^8 t and natural gas of 5000×10^8 m^3, with an average burial depth of over 7,300 m, and the characteristics of ultra-deep, ultra-high pressure and ultra-high temperature. The discovery of Shunbei Oilfield is a major breakthrough in domestic deep oil and gas exploration. Large-scale carbonate gas fields such as Puguang, Longgang and Gaoshiti have also been found in Sichuan Basin (Jizhong and Shengji 1993; Wenhai 1996; Yongsheng et al. 2007; Shugen et al. 2008; Longde et al. 2013; Guoqi et al. 2013; Jinhu et al. 2014). The main reservoir of Wubaiti Gasfield is composed of dolomite in the middle Huanglong Formation of Carboniferous strata (Ping 1998; Rongcai et al. 2014), with the maximum buried depth of 4595 m. It is a large-scale stratum-structure gas reservoir, with proved reserves of 587.11×10^8 m^3. The deepest gas-bearing structure in China is Laojunmiao Gasfieldin Sichuan Basin, with a depth of 7153.5–7175 m. The main reservoir of Jingbian Gasfield in Ordos Basin is Ma5 member of Ordovician

Majiagou Formation, part of which is buried below 3500 m, with the maximum buried depth of 3600 m. The exploration of deep oil and gas reservoirs in Bohai Bay Basin had begun in 1977. The geological reserves of 73×10^4 t^3 were obtained in well Ninggu 1, dolomite Buried Hill in Lining Oilfield, with the maximum buried depth of 5,300 m. At present, more than 100 deep oil and gas reservoirs have been found, and the proved geological oil reserves are more than 2×10^8 t. The deep strata of Songliao Basin mainly refer to the strata below Quantou Formation, with the buried depth more than 3000 m. For example, the buried depth of Changde Gasfield in the west slope zone of Xujiaweizi fault depression is nearly 3600 m. In recent years, a series of breakthroughs have been made in the exploration of deep volcanic natural gas in Songliao basin. The daily production of well Xushen1, well Weishen5, well Zhaoshen10 and well Wangshen1 in Xujiaweizi fault depression exceeds 10×10^4 m^3, which guides to a gasfield with reserves of more than 100 billion cubic meters.

1.2.2 Exploration Trend of Deep Oil and Gas in China

Nowadays, the oil and gas exploration onshore has been continuously developed to deep and ultra-deep strata. Major exploration breakthroughs have been made in the eastern region below 4500 m and the western region below 6000 m. The deep strata has become a major replacement field for onshore oil and gas exploration in China since the exploration depth extended 1,500–2,000 m down. Taking CNPC as an example, the average depth of exploratory wells increased from 2296 m in 2000 to 3190 m in 2018, with an increase of 894 m. The depth of exploratory wells in the east of China has been increased continuously, and has exceeded 6000 m, such as 6027 m in well Niudong 1 in 2011. While the depth of exploratory wells in the central and western regions of China has increased obviously, which has exceeded 8000 m, such as 8023 m in Keshen 7 well in 2010 and 8060 m in Wutan 1 well in 2018. Since 2000, the proportion of increased reserves of deep oil and gas of CNPC has been on the rise. Before 2010, the proportion of deep oil reserves averaged 6.8%, and since 2010, the proportion has increased to 13.4%. Before 2010, the proportion of deep nature gas reserves averaged 31.6%, and the proportion has increased to 57.2% since 2010 (Fig. 1.4).

At present, domestic onshore deep oil and gas exploration is in the period of breakthrough and discovery, more deep oil and gas fields are yet to be explored and developed. Taking CNPC as example, 14 important discoveries have been made in recent deep oil and gas exploration, including 5 oil discoveries and 9 natural gas discoveries. In the exploration area of carbonate rocks, two reserve areas with a scale of more than 5×10^8 t were found in Tabei area and Tazhong area of Tarim Basin. A reserve area with a scale of more than 3000×10^8 m^3 was found in Longgang Reef Beach of Sichuan Basin. In the exploration area of clastic rocks, gas fields with a scale of trillion cubic meters were found in Kuqa area of Tarim Basin, Songliao Basin and deep potential strata of Qikou area in Bohai Bay Basin. In the exploration area of volcanic rocks, two gas fields with a scale of 100 billion cubic meters in Xushen area of Songliao

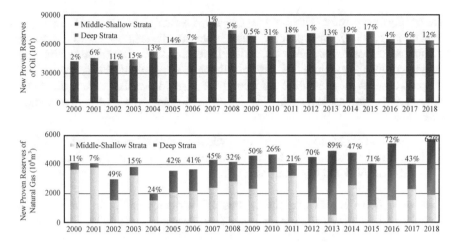

Fig. 1.4 Distribution of new proved reserves of oil and gas in China since 2000

Basin and Kelameili area of Junggar Basin, and an oilfield with a scale of 5000×10^4 t in Niudong area were found. However, there are abundant deep oil and gas resources in China, including deep oil resources (>4500 m) of 304.08×10^8 t, accounting for 28% of the total national oil resources; deep natural gas resources (>4500 m) of 29.12×10^{12} m³, accounting for 52% of the total national natural gas resources. Within China's petroleum mining claims, deep oil resources amount to 144.37×10^8 t, accounting for 29% of China's total oil resources, and the deep natural gas resources amount to 26.73×10^{12} m³, accounting for 56% of China's total oil resources. In terms of remaining hydrocarbon resources, the remaining oil resources in deep strata are 128.09×10^8 t, and the proved rate is only 11.3%, while the remaining gas resources in deep strata are 23.11×10^{12} m³, and the proved rate is 13.5%, which is far lower than that in middle-shallow strata. The above studies indicate that the deep strata may be an important area for reserve and production increase.

1.2.3 Geological Characteristics of Deep Oil and Gas in China

China is rich in deep oil and gas resources and has great exploration potential. However, the domestic petroliferous basins have their own characteristics.They are mostly developed on the small cratonic blocks, with small scale and poor stability, and they can creat the conditions for the formation and accumulation of oil and gas resources, as the sedimentary and structural differentiation developed in platform through the strong reforming and destruction caused by the sedimentary environ-ment with great lateral change. At the same time, because of the early formation of craton plate and multistage of tectonic evolution, the domestic petroliferous basins are mainly composed of superimposed basins, developing two sets of sedimentary structures (Fig. 1.5). The upper structural strata (middle-shallow strata) are mainly composed of continental sediments, and the lower structural strata (deep to ultra-deep strata) are mainly composed of marine sediments. The geological characteristics of deep oil and gas are quite different from those in middle-shallow strata, due to the deep buried depth, ancient strata, long history of diagenesis, high degree of thermal evolution.

(1)   Two kinds of hydrocarbon source kitchens capable of providing hydrocarbons on a large scale: conventional hydrocarbon source kicthens and gas source kitchens of liquid hydrocarbon cracking

The conventional source kitchens are widely distributed and multi strata, mainly including mudstones, carbonate rocks and coal-measure source rocks. For example, five sets of source rocks with an area of 26×10^4 km^2 have been discovered in the Paleozoic strata of Tarim Basin. The gas source kitchens of liquid hydrocarbon cracking include liquid hydrocarbon remained in paleo-oil reservoir and source rocks. The results show that the gas-generating potential of cracking liquid hydrocarbon is 2–4 times of the same amount of kerogen. The main gas-generating period of cracking liquid hydrocarbon is late, and R_0 is generally in the range of 1.6–3.2%, which is favorable for natural gas accumulation and preservation in the late stage.

(2)   Hydrocarbon-generating mechanisms in deep oil and gas exploration and breakthrough of "dead line" in deep oil and gas exploration

The traditional mechanism of hydrocarbon generation is based on the process of the primary pyrolysis of kerogen and hydrocarbon generation, without considering the influence of pressure to the process of secondary pyrolysis into gas and catalysis after the first pyrolysis of organic matters into oil. The research of "hydrocarbon generation hysteresis mechanism in deep ultra-high pressure environment" shows that (Cha Ming et al. 2002), large-scale hydrocarbon generation in deep strata is still possible in the late hydrocarbon generation period under the double effects of temperature and pressure, because the abnormally high pressure will inhibit the thermal evolution of organic matters, and then inhibit the formation and decomposition of hydrocarbons, delaying the role of hydrocarbon generation. The research of "successive generation mechanism of residual hydrocarbon" shows that, the characteristics of strong ability and large amount of hydrocarbon generation are revealed in the dispersed soluble organic matters out of the source rocks (residual hydrocarbon) (Figs. 1.6 and 1.7). In the ancient marine strata of Ordos, Sichuan and Tarim Basin, the residual hydrocarbons are widely distributed in various strata due to the multi-stage structural superimposition. The experimental analysis of Wenzhi et al. (2005, 2011) show that the gas-generation ability of dispersed soluble organic matter is about 2–4 times as much as kerogen. The theory also reveals that the thermal evolution of dispersed soluble organic matter out of source is later than that of organic matter inner source, which results in the succession of thermal evolution of dispersed soluble organic matters between out of source and inner source. The mechanism

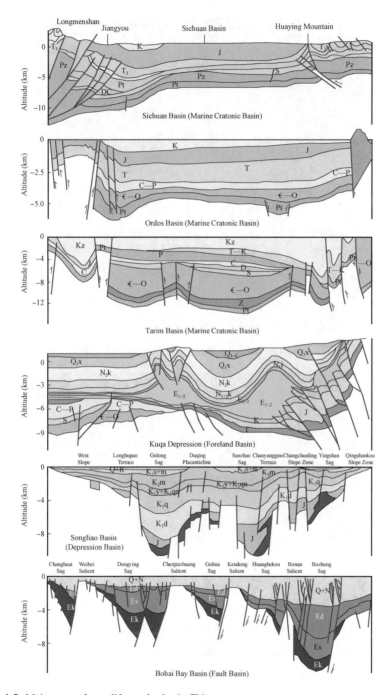

Fig. 1.5 Main types of petroliferous basins in China

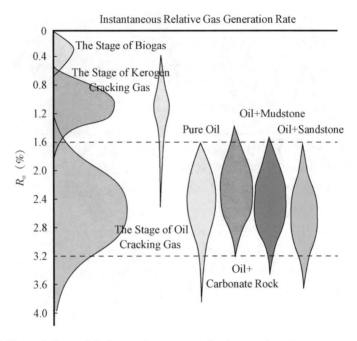

Fig. 1.6   The evolution model of successive gas generation by organic matter

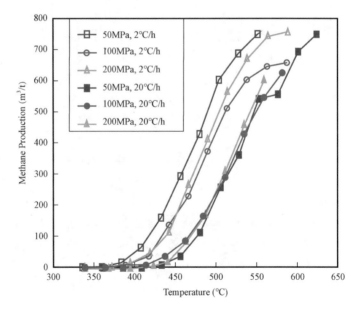

Fig. 1.7   Comparison of gas production quantity by crude oil cracking under different pressure conditions

of "organic-inorganic compound hydrocarbon generation" shows that the hydrogen in deep fluid can significantly improve the hydrocarbon generation efficiency of source rocks. Under the influence of deep fluids, poor-hydrogen source rocks (such as source rocks with high maturity) widely distributed in the deep basins still have good hydrocarbon-generating potential, and lay a solid material base of hydrocarbon generation of kerogen. The development of various hydrocarbon generation mechanisms have provided a theoretical basis for the formation of deep oil and gas reservoirs by deepening the understanding of deep hydrocarbon resources and potential.

According to the traditional theory of hydrocarbon geology, the formation of oil and natural gas is closely related to the temperature. The researches of Bjorkum from Norway shows that the strata with the temperature from 60 to 120 °C are the "golden zone" for oil and gas exploration, with 90% of global oil and gas stored there. However, in recent years, the extinction boundary between oil and natural gas is extened by the breakthroughs of "liquid window", "gas window" and "golden zone" through the exploration of deep oil and gas (Table 1.2). Through the statistics (Guangya et al. 2015) of 428 oil and gas fields with production strata deeper than 5000 m in the world, it can be seen that, there are 83 oil and gas fields with temperature greater than 150 °C, with the highest temperature of 375 °C. And there are 198 oil and gas fields with pressure coefficient greater than 1.2, including 60 oilfields and 138 gasfields, and the highest pressure coefficient is 2.8.

(3)   The lithology and types of deep reservoirs are various, and the formation mechanisms are diverse

The properties in reservoirs are generally tight due to the increase of overburden pressure and diagenetic cementation under deep geological conditions. Recent deep oil and gas exploration shows that, the effective porosity of the reservoirs can still reach 5–10% or even higher when the buried depth of the reservoirs exceeds the traditional "dead line" of 4500 m. Recent researches show that carbonate, clastic and volcanic reservoirs can be developed in deep and ultra-deep onshore strata in China, which can be the condition for oil and gas accumulation. Taking carbonate rocks as an example, large-scale effective reservoirs can still be developed in deep and ultra-deep carbonate rocks, which are affected by bedding karstification, interlayer karstification and dolomitization such as burial dolomitization and hydrothermal dolomization. For example, the Yingshan Formation of Ordovician, which buried below 6000 m in the slope area of the Tabei Uplift, Tarim Basin, developed widely in the layered karst reservoirs in the slope and the lower part under the bedding karstification, with the reservoir area exceeding 1.5×10^4 km^2. According to the relationship between physical properties and depth of 5860 samples in carbonate reservoirs of two basins, Tarim Basin and Sichuan Basin, the samples whose porosity is more than 5% still account for 35% of the statistical samples, and the maximum porosity is 22%. There are many types of reservoirs in deep strata, including carbonate reservoir, clastic reservoir, volcanic reservoir and metamorphic reservoir. The deep reservoir space is diversified, and the heterogeneity of reservoirs is caused by various types of reservoir space, such as pores, karst caves and fractures. Overall, the deep reservoirs have

Table. 1.2 Comparison of "oil (gas) window" and "dead line" in reservoir between traditional theory and deep exploration practice

Comparison	Liquid window		Gas window		Oil extinction boundary		Natural gas extinction boundary		Bottom depth of reservoir (km)
	Ro (%)	Temperature (°C)	Ro (%)	Temperature (°C)	Ro (%)	Temperature (°C)	Ro (%)	Temperature (°C)	
Oil and gas in traditional middle-shallow strata	0.5–1.35	60–150	1.35–3.6	180–250	<5	150–170	<10	250–375	<4.5
Oil and gas in deep strata	Upper limit >1.35 Up to7–8	150–295	Upper limit >4	>300 Experiemntal results <800	>5	>170 Experiemntal results <400	>10	>375	>4.5

scale, such as beach reservoirs in Longwangmiao Formation of Gaoshiti-Moxi area, Sichuan Basin, with the area of 3000 km^2, the clastic reservoirs in Kuqa with the area of 1856 km^2, the volcanic reservoirs in Karameili Gasfield with the area of 5000 km^2.

The deep high-quality reservoirs can be divided into two types: primary-preserved high-quality reservoir and secondary-formed high-quality reservoir. The former is developed with deep overpressure protection, deep-water turbidite sandstone and reef beach, and the latter is developed with dolomitization of carbonate rocks, dissolution and fracture.

① Deep overpressure protection. Deep overpressure can be formed by rapid settle-ment of basins, rapid filling of sediments, undercompaction of early strata, and stress modification by tectonic compression in late period. The deep overpres-sure can maintain more primary pores in reservoirs by effectively reducing the mechanical compaction. At the same time, overpressure can also weaken the chemical cementation caused by deep fluid flow.

② Deep-water turbidite sandstone. The high-quality sandstones characterized by late formation, weak diagenesis and good properties are formed by the rapid accumulation of high-quality sandstones supplied to passive continental margin by ancient rivers through overall transportion of deep-water gravity flow. The sandstones are rootless tongue-like bodies in plane, with thick and wide distribution, massive and dllipsoidal in section.

③ Reef beach reservoir.

④ Dolomitization. The secondary carbonate reservoirs are formed by dolomization of limestone through the great improvement of porosity and brittleness in the reservoir. Recent exploration practice reveals that, deep high-quality reservoirs can be formed under the burial dolomitization and hydrothermal dolomitiza-tion. The depth of effective dolomite reservoir formed by dolomitization in Tarim basin can up to 8000 m, such as Well Tashen 1, with the buried depth of 8104 m and the porosity of dolomite reservoir can still reach 5.2%. Affected by the invasion caused by deep Mg-rich fluid through the regional faults and fractures, the dolomite reservoirs are formed by hydrothermal dolomitization, and distribued in grid pattern. This kind of dolomite reservoir encountered in Well Gu9 of Tarim Basin from 6218 to 6314 m has a porosity of 12.6%.

⑤ Dissolution. The deep dissolution is mainly reflected in the bedding karstifi-cation and interbedded karstification of carbonate rocks. Stratiform large-area reservoirs of high-quality are formed when the bedding karstification is driven by head difference and atmospheric water flows down the slope with a depth of dissolution of hundreds to thousands of meters., Experiments in recent years show that sandstone can be dissolved quickly under high temperature and high pressure, and the dissolution rate of sandstone will increase by 2–3 times when the temperature is more than 150 °C. The research of the relationship between the porosity and acoustic wave in Cretaceous strata of Kuqa area shows that, there is still a high porosity at a depth of about 5800 m, which is presumed to be the contribution of sandstone dissolution. In addition, the reservoir can be greatly improved by dissolution as the properties of volcanic reservoir are less affected by compaction. For example, in the Carboniferous strata of northern Xinjiang, the shorter the distance from the strata to unconformity top boundary, the higher the porosity is.

⑥   Fracture. Whether it is carbonate rock, clastic rock, or igneous rock, the common feature of deep reservoirs is the development of fractures, especially near the fault zone where the fractures are densely distributed and form a network. According to the deep wells in Tarim Basin, the fractured reservoirs have gradually become the subject of the carbonate reservoirs from the buried depth of 5,700 m. Below 7000 m, fractured reservoirs have become the major part of reservoirs. For sandstone in Kuqa area, the porosity of reservoirs buried below 7000 m is generally less than 5–8%, and the permeability is 1–100 mD. As an effective reservoir with high gas flow, the tight sandstone is closely related to the improving effect of fractures on reservoir.

(4)   Strong diagenesis, poor reservoir connectivity, complicated distribution and relationship of oil and water

The accumulation model of carbonate reservoir is formed as "one cave, one reservoir" due to the coexistence of fractures and caves with different scales caused by the strong diagenesis and poor connectivity of deep reservoirs. The thermal evolution process and hydrocarbon generation of organic matters in deep strata are very different from those in middle-shallow strata due to the special temperature-pressure conditions caused by a complicated and changeable evolution process of temperature and pressure. Firstly, the distribution relationship between oil and water is complex with the development of edge water and bottom water, and the interlayer water, oil and gas coexist with each other. Secondly, the fluid phase behavior of hydrocarbon is complex. According to the present exploration practice, there are still large-scale reserves in deep strata despite of the complicated phase behavior of hydrocarbon under the condition of high temperature and high pressure. For example, in the Moxi area of Sichuan Basin, a large-scale and high-pressure equipped gasfield (141.4 and 1.65 °C) was found in the Longwangmiao Formation, Cambrian, while the normal crude oil was found in the well Jinyue 1 on the south slope of Tabei area, Tarim Basin, at buried depth of 7200 m (>180 °C) (Caineng et al. 2014).

(5)   Through multi-stage hydrocarbon supply, multi-stage hydrocarbon filling and reservoir reformation, the reservoirs are distributed in clsuter. The complex processes of hydrocarbon generation and accumulation are caused by large-area, long-history of accumulation evolution, and multi-period tectonic movement.

Compared with foreign areas, petroliferous basins in China have extremely complex evolutionary history. Generally, the evolutionary history can be divided into three stages: the first stage is the formation of platform basin and aulacogen, which shows the dispersion; the second is the formation of craton depression, which shows the total settlement; and the third is the opposition between basin and mountain in pseudo-foreland basin, which shows the extrusion. After the multi-stage tectonic changes mentioned above, most of the deep strata have been reformed, some even changed beyond recognition. Accordingly, the processes and mechanisms of migration and accumulation in deep strata are obviously different from those in middle-shallow strata, as multi-stage migration, accumulation, adjustment and transformation happened in deep strata.

1.2.4   Problems and Challenges of Deep Oil and Gas Exploration in China

In recent years, some important achievements have been made in deep oil and gas exploration in China. However, the traditional petroleum geology theory and technology system can not be the effective guidance and influence the selection of major exploration replacement fields due to the many challenges in deep oil and gas exploration, the effective development and engineering technology caused by the differences of hydrocarbon generation, reservoir, accumulation environment and mechanism between the geology in deep strata and middle-shallow strata.

(1)   Identification and hydrocarbon generation mechanism of ancient source rocks (Precambrian)

The exploration of hydrocarbon generation mechanism of source rocks under high-temperature and high-pressure condition has been carried out by predecessors. However, the exploration field and direction are still restricted because of a big gap in the identification of ancient hydrocarbon source rocks, hydrocarbon generation mechanism and the gas generation potential of dispersed liquid hydrocarbon cracking. The Precambrian is one of the most important hydrocarbon-bearing strata in the world, for example, the major oilfields in the east Siberia and Amman are found in the Precambrian reservoirs. In China, the three major Neoproterozoic-Cambrian ancient landmasses, north China landmass, Yangtze landmass, and Tarim landmass, are developed in the favorable tectonic environment where the source rocks are developed, and large-scale reserves have been found in the Sinian and Cambrian strata in Sichuan Basin. However, the formation environment, material composition and spatial distribution of these ancient source rocks are still unclear and need futher study. In addition, the burial depth and maturity of ancient source rocks in China are generally on the high side, and the hydrocarbon generation mechanism of high-overmature source rocks under high temperature and high pressure is not clear, besides dispersed liquid hydrocarbon are lack of effective means for the evolution of gas generation potential. As a result, the evaluation of deep source rocks and the potential understanding of resources are affected, and the material composition, hydrocarbon generation mechanism and accumulation potential of ancient source rocks are not clear.

(2)   Formation and preservation mechanism of deep reservoirs

The complex origins of deep reservoirs are caused by the superposition of multi-
stage diagenesis. The selections of favorable areas and targets for exploration are
restricted by the misunderstanding of mechanisms in reservoir diagenesis, pore development, and preservation of large-scale effective reservoir, as well as multiple stages
of superimposed transformation, the main controlling factors, distribution prediction
and distribution law in large-scale high-quality reservoir under high temperature and
high pressure.

(3)   Accumulation in cross-tectonic period and hydrocarbon distribution law of oil
       and gas

Under the condition of high temperature and high pressure, the section of exploration targets and breakthrough point are restricted by the misunderstanding of accumulation mechanism, accumulation process and hydrocarbon distribution in cross-
tectonic period. The evaluation of exploration target can not be accomplished due
to the lack of researches in hydrocarbon filling mechanism under high temperature
and high pressure, the accumulation mechanism composed of gypsum salt rocks
and carbonate rocks, the hydrocarbon accumulation model and the distribution of
large oil and gas fields. Most of the sedimentary basins in China are superimposed
basins, with complex geological conditions, which caused by the transformation of
multi-stage tectonic activities. The further researches on the mechanism of oil and
gas accumulation and the main controlling factors of large-scale oil and gas fields
in superimposed basins are still needed, especially in the fields of: ① the accumulation power and hydrocarbon filling mechanism in the high-temperature and high-
pressure closed system, including the fluid phase behavior and filling mechanism in
the high-temperature and high-pressure environment, and the oil and gas migration
and accumulation power in the high-temperature and high-pressure closed environment; ② the accumulation mechanism and accumulation process of combination
of gypsum salt rocks and carbonate rocks, including the influence of hydrocarbon
formation and reservoir formation in gypsum salt environment, and the accumulation mechanism and process of the combination; ③ the process and efficiency of
oil-gas accumulation in the cross-tectonic period, including the relationship between
tectonic evolution and oil-gas enrichment, the mechanism and process of oil-gas
accumulation in the cross-tectonic period; ④ the formation and distribution of large
oil and gas fields under high-temperature and high-pressure, mainly including the
migration and accumulation mechanism of oil and gas, and accumulation mode in
broad and gentle structure, formation conditions and accumulation rules in deep large
oil and gas fields, evaluation and prediction technology of deep oil and gas accumulation areas. These problems make it difficult to select breakthrough areas and key
zones for deep exploration, as well as replacement areas and zones for large-scale
exploration.

     In view of the aforementioned key geological problems, considering exploration
research and latest developments of deep oil and gas in key onshore petroliferous

basins in China in recent years, such as Sichuan Basin, Tarim Basin and Ordos Basin, this book provides theoretical guidance and technical support for deep hydrocarbon exploration breakthrough and large-scale increasing reserves by analyzing and summarizing systematically the formation conditions, distribution laws and main controlling factors of deep oil and gas fields, and predicting the continental exploration prospect of deep oil and gas in China.

References

Aase, N.E., and O. Walderhaug. 2005. The effect of hydrocarbons on quartz cementation: Diagenesis in the Upper Jurassic sandstones of the Miller field, North Sea, revisited. *Petroleum Geoscience* 11 (3): 215–223.

Ajdukiewicz, J.M., P.H. Nicholson, and W.L. Esch. 2010. Prediction of deep reservoir quality from early diagenetic process models in the Jurassic eolian Norphlet Formation, Gulf of Mexico. *AAPG Bulletin* 94 (8): 1189–1227.

Caillet, G., N.C. Judge, N.P. Bramwell, et al. 1997. Overpressure and hydrocarbon trapping in the Chalk of the Norwegian Central Graben. *Petroleum Geoscience* 3: 33–42.

Caineng, Zou, Du Jinhu, Xu Chunchun, et al. 2014. Formation, distribution, resource potential and discovery of the Sinian-Cambrian giant gas field, Sichuan Basin, SW China. *Petroleum Exploration and Development* 41 (3): 278–293.

Cao, B.F., G.P. Bai, and Y.F. Wang. 2013. More attention recommended for global deep reservoirs. *Oil & Gas Journal* 111 (9): 78–85.

Chunchun, Xu, Zou Weihong, Yang Yueming, et al. 2017. Status and prospects of exploration and exploitation of the deep oil & gas resources onshore China. *Natural Gas Geoscience* 28 (8): 1139–1153.

Ehrenberg, S.N., P.H. Nadeau, and O. Steen. 2008. A megascale view of reservoir quality in producing sandstones from the offshore Gulf of Mexico. *AAPG Bulletin* 92 (2): 145–164.

Fang, Hao, Zou Huayao, Ni Jianhua, et al. 2002. Evolution of over pressured systems in sedimentary basins and conditions for deep oil/gas accumulation. *Earth science Journal of China University of Geosciences* (in Chinese) 27 (5): 610–615.

Guangming, Zhai, Wang Shihong, and He Wenyuan. 2012. Hotspot trend and enlightenment of global ten-year hydrocarbon exploration. *Acta Petrolei Sinica* 33 (Supplement 1): 14–19.

Guangya, Zhang, Ma Feng, Liang Yingbo, et al. 2015. Domain a theory-technology progress of global deep oil & gas exploration. *Acta Petrolei Sinica* 36 (9): 1156–1166.

Guangyou, Zhu, Zhang Shuichang, Zhang Bin, et al. 2010. Reservoir types of marine carbonates and their accumulation model in western and central China. *Acta Petrolei Sinica* 31 (6): 871–878.

Guoping, Bai, and Cao Binfeng. 2014. Characteristics and distribution patterns of deep petroleum accumulatons in the world. *Oil & Gas Geology* (in Chinese) 35 (1): 19-25.

Guoqi, Wei, Shen Ping, Yang Wei, et al. 2013. Formation conditions and exploration prospects of Sinian large gasfields, Sichuan Basin. *Petroleum Exploration and Development* 40 (2): 129–138.

Hansheng, Q., and Li Feng. 2000. The depth petroleum geology and exploration. *Explorationist* 5 (4): 10–15.

Haiqing He, Wang Zhaoyun, and Han Guilong. 1998. Deep zone reservoir type and oil gas distribution pattern in the Bohai Gulf Basin. *Petroleum Exploration and Development* (in Chinese) 25 (3): 6–9.

Hunt, J. 1990. Generation and Migration of petroleum from abnormally pressured fluid compartments. *AAPG Bulletin* 74 (1): 1–12.

Jianghai, Li, Wang Honghao, and Li Weibo. 2014. Discussion on global tectonics evolution from plate reconstruction in Phanerozoic. *Acta Petrolei Sinica* 35 (2): 207–218.

Jiarui, Feng, Gao Zhiyong, Cui Jinggang, et al. 2016. The exploration status and research advances of deep and ultra-deep clastic reservoirs. *Advances in Earth Science* 31 (7): 718–736.

Jincai, Tuo, Wang Xianbin, Zhou Shixin, et al. 1999. Current situation about the research in oil and gas contained in deep formations and related progress. *Natural Gas Geoscience* 10 (6): 1–8.

Jinghong, Wang, Jin Jiuqiang, Zhu Rukai, et al. 2011. Characters and distribution patterns of effective reservoirs in the Carboniferous volcanic weathering crust in Northern Xinjiang. *Acta Petrolei Sinica* 32 (5): 757–766.

Jinhu, Du, Zou Caineng, Xu Chunchun, et al. 2014. Theoretical and technical innovations in strategic discovery of a giant gas field in Cambrian Longwangmiao Formation of Central Sichuan Paleo-uplift, Sichuan Basin. *Petroleum Exploration and Development* 41 (3): 268–277.

Jinxing, Dai, Ni Yunyan, Zhou Qinghua, et al. 2008. Significances of studies on natural gas geology and geochemistry for natural gas industry in China. *Petroleum Exploration and Development* 35 (5): 513–525.

Jizhong, Huang, and Chen Shengji. 1993. Hydrocarbon source analysis of Sinian gas reservoir forming conditions of the Sichuan Basin: A case history of Weiyuan gas field. *Natural Gas Geoscience* 4: 16–20.

Juan, Huang, Ye Deliao, and Han Yu. 2016. Petroleum geology features and accumulation controls for ultra-deep oil and gas reservoirs. *Petroleum Geology & Experiment* 38 (5): 635–640.

Lerche, I., and A. Lowrie. 1992. Quantitative models for the influence of salt-associated thermal anomalies on hydrocarbon generation, Northern Gulf of Mexico Continental Margin. *Gulf Coast Association of Geological Societies Transactions* 42: 213–225.

Longde, Sun, Zou Caineng, Zhu Rukai, et al. 2013. Formation, distribution and potential of deep hydrocarbon resources in China. *Petroleum Exploration and Development* 40 (6): 641–649.

Ma, Yongsheng, Guo Tonglou, Zhao Xuefeng, et al. 2007. Formation mechanism of deep-buried quality carbonate reservoir in Puguang Gasfield. *Science in China* 37 (supplement H): 43–52.

Ministry of Land and Resources. 2005. Regulation of Petroleum Reserves Estimation. DZ/T 0217-2005. Beijing: China Standard Publishing House.

O'Brien, J.J., and I. Lerche. 1988. Impact of heat flux anomalies around salt diapirs and salt sheets in the Gulf Coast on hydrocarbon maturity: Models and observations. *Gulf Coast Association of Geological Societies Transactions* 38: 231–243.

Ping, Shen, and Xu Renfen. 1998. Reservoir formation conditions of Wubaiti Gasfield in east Sichuan and its efficient exploration experiences. *Natural Gas Industry* 18 (6): 5–9.

Rongcai, Zheng, Li Ke, Ma Qike, et al. 2014. Diagenetic facies of carbonate rock reservoirs in Huanglong formation of Wubaiti gas field, East Sichuan, China. *Journal of Chengdu University of Technology* 41 (4): 401–412.

Shugen, Liu, Ma Yongsheng, Sun Wei, et al. 2008. Studying on the differences of Sinian natural gas pools between Weiyuan gas field and Ziyang gas-bearing area, Sichuan Basin. *Acta Geological Sinica* 82 (3): 328–337.

Tao, Wang. 2002. *Deep basin gas fields in China*. Beijing: Petroleum Industry Press.

Tissot, B.P., and D.H. Welte. 1978. *Petroleum formation and occurrence: A new approach to oil and gas exploration*, 185–188. New York: Springer.

Wenhai, Song. 1996. Research on reservoir-formed conditions of large-medium gas fields of Leshan-Longnvsi Palaeohigh. *Natural Gas Industry* 16 (supplement 1): 13–26.

Wenhui, Liu, Chen Mengjin, Guan Ping, et al. 2009. *Natural gas hydrocarbon generation, accumulation, three element geochemical tracer system and practice*, 119–126. Beijing: Science Press.

Wenhui, Liu, Wang Jie, Tenger, et al. 2012. Multiple hydrocarbon generation of marine strata and its tracer technique in China. *Acta Petrolei Sinica* 33 (S1): 115–125.

Wenzhi, Zhao, Zhang Guangya, and Wang Hongjun. 2005. New achievements of petroleum geology theory and its significances on expanding oil and gas exploration fields. *Acta Petrolei Sinica* 26 (1): 1–7.

Wenzhi, Zhao, Wang Zecheng, Zhang Shuichang, et al. 2007. Analysis on forming conditions of deep marine reservoirs and their concentration belts in superimposed basins in China. *Chinese Science Bulletin* 52 (Supplement I): 9–18.

Wenzhi, Zhao, Wang Zhaoyun, Wang Hongjun, et al. 2011. Further discussion on the connotation and significance of the natural gas relaying generation model from organic materials. *Petroleum Exploration and Development* 38 (2): 129–135.

Wenzhi, Zhao, Shen Anjiang, Hu Suyun, et al. 2012. Geological conditions and distributional features of large-scale carbonate reservoirs onshore China. *Petroleum Exploration and Development* 39 (1): 1–12.

Wenzhi, Z., Hu Suyun, Liu Wei, et al. 2014. Petroleum geological features and exploration prospect in deep marine carbonate strata onshore China: Further discussion. *Natural Gas Industry* 34 (4): 1–9.

Wilkinson, M., D. Darby, R.S. Haszeldine, et al. 1997. Secondary porosity generation during deep burial associated with overpressure leak-off, Fulmar Formation, U.K. Central Graben. *AAPG Bulletin* 81: 803–813.

Xiaoguang, Tong, Zhang Guangya, Wang Zhaoming, et al. 2014. Global oil and gas potential and distribution. *Earth Science Frontiers* 2 (3): 1–9.

Xiaorong, Luo, Zhang Likuan, Fu Xiaofei, et al. 2016. Advances in dynamics of petroleum migration and accumulation in deep basins. *Bulletin of Mineralogy, Petrology and Geochemistry* 35 (5): 876–889.

Xiongqi, Pang. 2010. Key challenges and research methods of petroleum exploration in the deep of super imposed basins in western China. *Oil & Gas Geology* 31 (5): 517–534.

Yuzhu, Kang. 2008. The development of the gas geochemistry characteristics of the gas fields in New China with more than 100 billion cubic meters of reserves based on the Paleozoic marine carbonate accumulation theory in China. *Marine Origin Petroleum Geology* 13 (4): 8–11.

Yu, Wang, Su Jin, Wang Kai, et al. 2012. Distribution and accumulation of global deep oil and gas. *Natural Gas Geoscience* 23 (3): 526–534.

Zecheng, Wang, Zhao Wenzhi, Zhang Lin, et al. 2002. *Tectonic sequence and natural gas exploration in Sichuan basin*, 1–287. Beijing: Geological Publishing House.

Zhaoyun, Wang, Zhao Wenzhi, Zhang Shuichang, et al. 2009. Origin of deep marine gas and oil cracking gas potential of Paleozoic source rocks in the Tarim Basin. *Acta Sedimentologica Sinica* 27 (1): 153–163.

Zhengzhang, Zhao, Du Jinhu, Zou Caineng, et al. 2011. Geological exploration theory for large oil and gas provinces and its significance. *Petroleum Exploration and Development* 38(5): 513–522.

Zhijun, Jin. 2005. Particularity of petroleum exploration on marine carbonate strata in China sedimentary basins. *Earth Sceinec Frontiers* 12 (3): 15–22.

Zhijun, Jin, Hu Wenxuan, and Zhang Liuping. 2007. *Deep fluid activities and oil and gas accumulation effect* (in Chinese). Beijing: Science Press.

Zhiyi, Zhang. 2005. Renew exploration concept and open up new field of deep oil and gas. *Oil & Gas Geology* 26 (2): 193–196.

Zhongjian, Qiu, and Fang Hui. 2009. Surging of natural gas in China: A new journey of China's petroleum industry. *Natural Gas Industry* 29 (10): 1–4.

Chapter 2
Deep Source Rocks and Hydrocarbon Generation Mechanism

It is challenging to evaluate the scale and effectiveness of hydrocarbon kitchens due to the deep buried depth of source rocks in deep large oil and gas fields. Previous studies have always focused on the effectiveness of muddy source rocks, paleo-reservoir and generation potential of residual hydrocarbon inside the source. However, in recent years, the problems related to deep source rocks, such as the development model and generation mechanism in saline environment, the material composition, development environment and generation mechanism of Precambrian source rocks, as well as the generation model of coal-formed gas have attracted attentions and discussions. It is clear that large-scale hydrocarbon generation of two types of hydrocarbon kitchens (conventional hydrocarbon kitchen and liquid hydrocarbon cracking kitchen) is the material basis for the formation of deep large oil and gas fields.

2.1 Types and Characteristics of Deep Hydrocarbon Kitchens

Two types of hydrocarbon kitchens are generally developed in the deep marine strata, one is the conventional marine hydrocarbon kitchen, which generated large amount of hydrocarbon resources with long time and complete evolution, experiencing "oil and gas generation" peaks, with the characteristics of early oil generation and late gas generation. The other is the gas kitchen formed by liquid hydrocarbons with different occurrence states, producing gas on a large scale in the high-over mature stage. These two kinds of source rocks are the material sources of deep hydrocarbon.

© Petroleum Industry Press 2021
S. Hu and T. Wang, *Deep-Buried Large Hydrocarbon Fields Onshore China: Formation and Distribution*, https://doi.org/10.1007/978-981-16-2285-4_2

2.1.1 Conventional Hydrocarbon Kitchen

The conventional hydrocarbon kitchens are generally developed in argillaceous rocks, carbonate rocks and coal-bearing source rocks, which are distributed. There are five sets of hydrocarbon kitchens in Tarim Basin, which are mainly composed of mudstone and argillaceous limestone. TOC ranges from 1.24 to 5.52%, with an average of 1.45%. The total thickness of hydrocarbon kitchens ranges from 250 to 750 m, with an area of 26×10^4 km^2. There are four sets of hydrocarbon kitchens in Sichuan Basin, which are mainly composed of mud shale and carbonaceous mudstone. TOC ranges from 1.04 to 6.52%, with an average of 2.58%. The total thickness of hydrocarbon kitchens ranges from 750 to 950 m, with an area of 19 $\times 10^4$ km^2. There are two sets of hydrocarbon kitchens in Ordos Basin, which are mainly composed of mud shale and argillaceous limestone. TOC ranges from 0.5 to 2.91%, with an average of 1.03%. The total thickness of hydrocarbon kitchens ranges from 20 to 160 m, with an area of 8×10^4 km^2. These source rocks with large thickness, wide distribution and high abundance of organic matters laid the material foundation of deep large oil and gas fields (Fig. 2.1).

Taking Sichuan Basin in northwest Yangzi Block (Digang et al. 2008) as a case study, this research comprehensively evaluates the distribution, thickness, evolution degree and hydrocarbon generation potential of source rocks of Cambrian, Doushantuo Formation of Sinian and Datangpo Formation of Nanhua System. The basement of Sichuan Basin, formed in the Jinning Movement, is composed of crystalline basement from Archean to Paleoproterozoic, Mesoproterozoic folded basement and Neoproterozoic Lower Sinian transitional basement. Under the influence

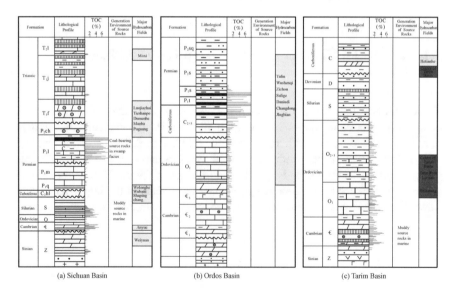

(a) Sichuan Basin   (b) Ordos Basin   (c) Tarim Basin

Fig. 2.1 Histogram of the main source rocks distribution in the three major marine basins of China

of global extension in Neoproterozoic, the rift and the passive continental margin are formed by the breakup of South China Plate and the breakup and subsidence of the Yangtze Block margin. Rifts with different scales are formed in the Central Sichuan block, and three sets of high-quality source rocks are developed, including the Dengying, Maidiping and Qiongzhusi Formations, with the total thickness of 300–450 m, the organic carbon content of 0.5–8.49% and an average value of 1.959 (409) samples. The environment is dominated by deep-water reducing environment from Late Sinian to Early Cambrian. The abundance of organic matters in sediments is high, as TOC of Qiongzhusi Formation in Cambrianin in Well Gaoshi 17 is 0.37–6.0%, with an average value of 2.17%, and the TOC of Maidiping Formation is 0.70–4.0%, with an average value of 1.67% (Fig. 2.2).

A large-scale transgression occurred during the sedimentary period in Dengying Formation Member III of Sinian, but the rift is still in the early stage. Affected by the development scale and the sediment supply, the thickness of the northern rift is relatively thick, generally from 20 to 30 m, and the mudstone of the Gaoshiti-Moxi area is relatively thin, generally from 10 to 30 m, with a limited distribution. From Yibin in the southern basin to Suijiang, the mudstone of Dengying Formation Member III is also developed with the thickness of 5–10 m (Fig. 2.3). Outside the rift area, the thickness of Dengying Formation Member III is less than 2 m. The thickness difference between inside and outside of the rift area is 2–10 times. The transgression of the Early Cambrian results in the deposition and filling of the black shale in Maidiping and Qiongzhusi Formations inside the craton rifting area. The source rocks of Maidiping Formation are mainly distributed inside the rifting, with the thickness of 50–100 m, while the surrounding area is only 1–5 m. The difference between them is more than 10 times (Fig. 2.3). The distribution of muddy source rocks in Qiongzhusi Formation are obviously controlle by rifting. The thickness of source rocks along the rift direction is the largest, with the thickness of 300–350 m, and the thickness on both sides of the depression is obviously thin. The thickness of source rocks in Qiongzhusi Formation of main rifting is 3–5 times of that in adjacent areas (Fig. 2.3). Generally speaking, the development of source rocks is controlled by the development of rifting trough. The source rocks of Doushantuo Formation in Yangtze Block are mainly distributed in the Middle-Lower Yangtze platform basin, slopes and deep-water basin, with the thickness of 30–379 m, and the general thickness of 60–149 m. The source rocks are mainly distributed in Yichang Gorge-Hefeng-Shimen-Tongren-Zunyi, Dexing-Kaihua-Ningguo and Liping-Sanjiang-Lingui-Quanzhou areas (Fig. 2.3). The thickness is generally more than 60 m with the thickest in Hefeng, up to 379 m. The thickness contours extend from southwest to northeast. In Liping-Sanjiang-Lingui-Fuzhou area, the thickness counters of source rocks extend from northwest to southeast. In addition, the thickness of source rocks in Ningqiang of Shaanxi and Wanyuan-Chengkou area is thicker, with 579 m in Hujiaba of Ningqiang, 310 m in Dazhu of Wanyuan, and 390 m in Mingyue of Chengkou. Except above areas, the thickness of source rocks in other areas is less than 60 m, while in Upper Yangtze is zero.

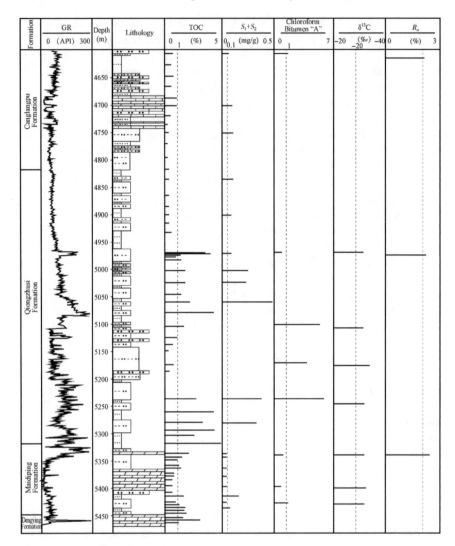

Fig. 2.2 Characteristics of Cambrian Source Rocks in Well Gaoshi 17

2.1.2 Source Kitchens by Liquid Hydrocarbon Cracking

The concept of source kitchen by liquid hydrocarbon cracking is proposed (Wenzhi et al. 2005a, b, 2006a, b, 2008, 2015) for at the deep burial, multi-stage tectonic evolutions and the hydrocarbon generation characteristics of "double peaking" in deep ancient source rocks in China. The source kitchen by liquid hydrocarbon cracking is not only the process of source material enrichment, but also an important condition for the large scale generation and accumulationof conventional and unconventional

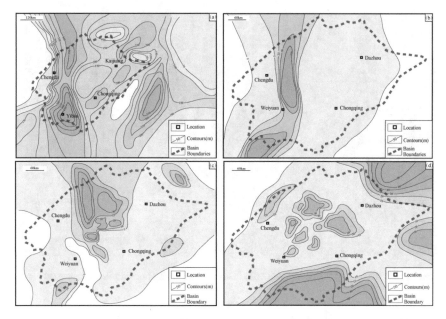

Fig. 2.3 Cantour map of thickness of four sets of source rocks in Sichuan Basin. **a** Thickness map of source rocks in Qiongzhusi Formation of Cambrian in Sichuan Basin; **b** Thickness map of source rocks in Maidiping Formation of Cambrian in Sichuan Basin; **c** Thickness map of source rocks in Dengying Formation Member III of Sinian in Sichuan Basin Basin; **d** Thickness map of source rocks in Doushantuo Formation of Sinian in Sichuan Basin

resources in late period. The source kitchen by liquid hydrocarbon cracking actually refers to the gas source kitchens formed by the late-period cracking of liquid hydrocarbon in three states under geological conditions (Fig. 2.4): the first is the cracking in the accumulated paleo-reservoir, that is, the liquid hydrocarbon accumulated outside the source; the secoucl is the halfway cracking of semi-accumulation and semi-dispersed "pan-reservoir", that is, the liquid hydrocarbon dispersed outside the source; the third is the late-period cracking in the residual hydrocarbon kitchen, that is, the liquid hydrocarbon dispersed inside the source. The dispersed liquid hydrocarbon outside the source is in the form of semi-accumulation and semi-dispersion in the process of accumulation due to the gentle structure and tight lithology resulting in low enrichment and unformed reservoir. The quantity and distribution ratio of liquid hydrocarbon in the three states are controlled by internal and external factors.

Oil cracking is an important way for gas generation in marine source rocks, and two necessary conditions must be satisfied, one is the formation of large amount oil in the stage of oil window, the other is a high temperature to satisfy the thermodynamic condition of oil cracking.

The material base of hydrocarbon kitchen is sufficient and gas generation efficiency is high. Compared with kerogen, the gas-cracking and generation period of liquid hydrocarbon is late and the amount of gas is large, which has been confirmed by

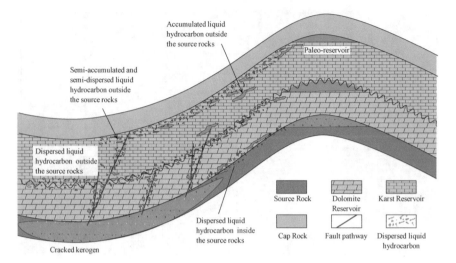

Fig. 2.4 Sketch map of the occurrence state in source kitchen by liquid hydrocarbon cracking

simulation experiment. In the system of high-temperature, high-pressure and semi-open, the scientific problems such as the material source of nature gas in high-over mature stage and the reservoir formation of residual hydrocarbons in source rocks are discussed. By the hydrocarbon expulsion simulation experiment in underground environment and the research of gas generation mechanism in different states, it is found that the cracking period of liquid hydrocarbon is later than that of kerogen, with the best time of 1.6–3.2% ofw Ro, and the amount of gas production is 2–4 times compared with the equal amount of kerogen. With huge accumulation potential (Fig. 2.5), the liquid hydrocarbon can be the major source of gas accumulation in deep carbonate rocks.

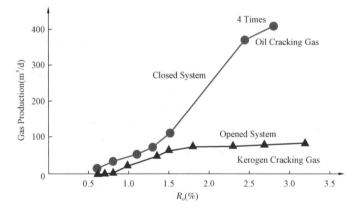

Fig. 2.5 Comparison of cracking gas yield between kerogen and liquid hydrocarbon

Changing the pressure will affect the thermal stability of cracking process. The cracking experiments under different temperature and pressure conditions show that the pressure influence on the cracking rate of crude oil is much weaker than that of temperature, and the response of cracking rate to pressure is not linear increase or decrease, but there is a maximum value (Behar et al. 1996). The oil cracking gas is the main part of natural gas in the marine basin. Because of the high temperature in the deep strata, it provides favorable geological conditions for the crude oil cracking. The greater burial depth, the greater amount of crude oil cracking gas resources may be. A large amount of oil cracking gas have been found recently in Sinian to Cambrian in the Sichuan Basin confirms this conclusion. The paleo-geothermal temperature in most areas of Paleozoic marine strata may exceed 230 °C because of the high paleo-geothermal gradient in Sichuan Basin (Fig. 2.6). Although the geothermal temperature in the most areas of Changxing-Feixianguan Formation is lower than 230 °C, the process of crude oil cracking into gas is still promoted. Because of the reduction of oil stability by TSR, the temperature of the complete pyrolysis of crude oil into gas was reduced to 120–160 °C. Therefore, the exploration is still dominated by natural gas as the hydrocarbon of Changxing-Feixianguan Group is in gaseous state.

The natural gas can be formed into reservoirs in late period because of the long time-window of liquid hydrocarbon cracking. Taking Anyue gasfield in Sichuan Basin as an example (Fig. 2.7), large-scale paleo-oil reservoirs are formed in Ziyang paleo-trap, Anyue paleo-trap, Weiyuan paleo-slope and Moxi-Gaoshiti area by migrating the oil and gas to the top of uplift zones and upper slopes when the organic

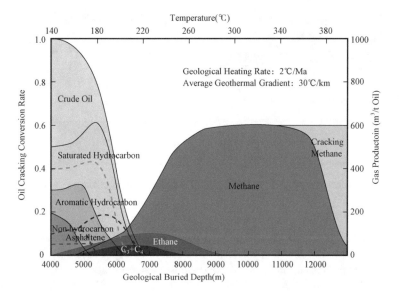

Fig. 2.6 Model of liquid hydrocarbon cracking and composition evolution in Sichuan Basin

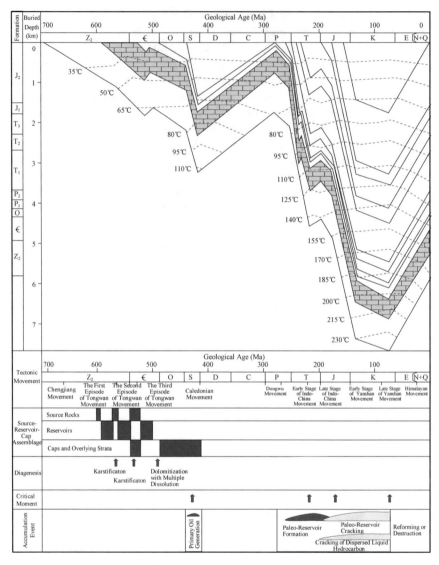

Fig. 2.7 Gas accumulation event in late period of Sinian-Cambrian in Anyue gas field, Sichuan Basin

matters of Sinian-Cambrian reached the peak stage of oil production in Permian-Middle Triassic. The Sinian-Cambrian formation was deeply buried by the cover of foreland basins with a thickness of 3000–5000 m since Late Triassic. The buried depth of the Sinian-Cambian formation is 7000–8000 m with the stratum temperature over 200 °C even at the axial zone of paleo-uplift in Leshan-Longnvsi. The paleo-reservoir and dispersed liquid hydrocarbon in Sinian-Cambrian cracked into gas and became important gas sources because of the high stratum temperature. The liquid hydrocarbon with semi-accumulation and semi-dispersion in slopes of central Sichuan paleo-uplift and paleo oil reservoirs in the core become the gas source kitchens for Anyue gas field. Three types of hydrocarbon sources contribute to the formation of large gas field, including cracking gas in paleo-oil reservoirs with the volume of 36.05×10^{12} m^3, the resource extent of $(1.8–3.6) \times 10^{12}$ m^3, accounting for 58–61% of total resources; cracking gas of dispersed liquid hydrocarbon in reservoirs with the volume of 76.59×10^{12} m^3, the resource extent of $(0.77–1.53) \times 10^{12}$ m^3, accounting for 25–26% of total resources; cracking gas of hydrocarbon retention in source rocks with the volume of 362.44×10^{12} m^3, the resource extent of $(0.36–1.08) \times 10^{12}$ m^3, accounting for 12–17% of total resources. It can be seen that the dispersed liquid hydrocarbons play an important role in the formation of large oil and gas fields. At the same time, the area with asphalt content over 3% is 5000 km^2 and the area with asphalt content of 1–3% is up to 6×10^4 km^2 in the central Sichuan paleo-uplist (Fig. 2.8). The asphalt is the residual product of paleo reservoir and the liquid hydrocarbon cracking with semi-accumulation and semi-dispersion (Fig. 2.9, and Chart I a–e).

According to the statistics of accumulation period in large marine gas fields in China (Table 2.1), the main accumulation periods of most gas fields or reservoirs are relatively late, with cracked liquid hydrocarbon as the main hydrocarbon source

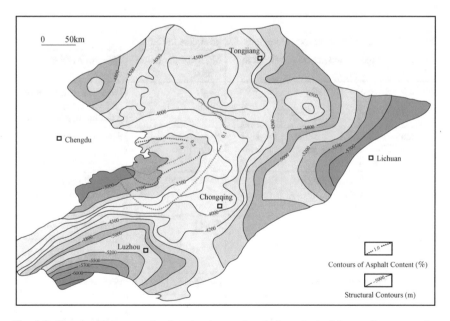

Fig. 2.8 Superposition map of paleo-structure and asphalt content of Lower Longwangmiao Formation before Jurassic in Sichuan Basin

Fig. 2.9 3D visualization of reservoir asphalt with semi-accumulation and semi-dispersion

Table 2.1 Accumulation periods of large marine gas fields in China

Basin	Large oil and gas fields	Reservoir	Main accumulation period	Types of hydrocarbon kitchen
Tarim	Tazhong, Gucheng	O	E—N	Paleo-oil accumulation and hydrocarbon retention inside source
Sichuan	Luojiazhai, Dukouhe, Puguang, Longgang, Yuanba,etc.	P—T	N—Q	Paleo-oil reservoir
	Tiandong, Datianchi, Wolonghe, Fuchengzhai	C	N—Q	Paleo-oil reservoir
	Moxi, Gaoshiti, Longnvsi, Hebaochang	$Z-\epsilon$	N—Q	Paleo-oil reservoir and liquid hydrocarbon with semi-accumulation and semi-dispersion
	Weiyuan, Ziyang	Z	N—Q	Paleo-oil reservoir
Ordos	Jingbian	O	K	Mainly coal source rocks

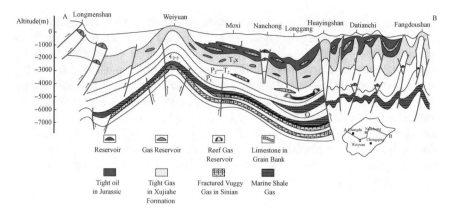

Fig. 2.10 Schematic diagram of "orderly symbiosis" of conventional and unconventional oil and gas reservoirs in Sinian-Silurian system, Sichuan Basin

kitchen. For example, most of the gas reservoirs in reef and beach in the northeast Sichuan Basin are cracked from paleo oil reservoirs, and the gas kitchens in Moxi-Hebaochang area are paleo oil reservoirs and liquid hydrocarbon with semi-accumulation and semi-dispersion.

Source kitchen of liquid hydrocarbon cracking is an important condition for orderly symbiosis of conventional-unconventional natural gas. Taking the Sinian-Cambrian Formation in Sichuan Basin as an example, cracking gas of liquid hydrocarbon can supply gas to the adjacent reservoirs, which not only formed the fracture-cavity gas reservoir of Dengying Formation in Sinian and pore-type conventional gas reservoir of Longwangmiao Formation in Cambrian (Fig. 2.10), but also formed unconventional shale gas reservoir in Qiongzhusi Formation in Cambrian and Longmaxi Formation in Silurian. The conventional-unconventional natural gas is symbiotic in space because of the large-scale nature gas resources formed from source rocks and reservoirs by the full supply of hydrocarbon from source kitchens of liquid hydrocarbon cracking.

2.2 Source Rocks and Hydrocarbon Generation Mechanism in Meso-Neoproterozoic

Oil and gas resources related to source rocks of Proterozoic have been found in many countries and regions around the world, such as Russia, Central Asia, North Africa and Australia. As of 2016, 80 oil and gas fields and 150 oil and gas reservoirs in the East Siberian Basin of Russia have been found, with proven oil reserves of 6.28×10^8 t, proven gas reserves of 2.02×10^{12} m^3, and oil equivalent of 22.36×10^8 t in hydrocarbon reserves. The Proven oil reserves of Neoproterozoic source rocks in Oman are 16.4×10^8 t. The geological reserves of the Neoproterozoic-Cambrian oilfields in Baghavala, India, are about 6.28×10^8 bbl.

Chinese scholars carried out Proterozoic hydrocarbon geology research as early as 1970s, and the discovery of Anyue gas field in Sinian-Cambrian of Sichuan Basin proves that there must be large potential for hydrocarbon exploration in Meso-Neoproterozoic. As a potential field for hydrocarbon exploration, one of the key factors in the exploration breakthrough is the "source". Taking microorganism as the starting point, this research provides experimental basis for the hydrocarbon resource evaluation and exploration selection of Proterozoic by studying Proterozoic paleo-ocean productivity, organic matter enrichment, hydrocarbon generation potential and discussing the microorganisms influence on the development of ancient source rocks through organic geochemistry, element geochemistry, biological fossil identification and microbial cultivation experiment.

2.2.1   Distribution of Source Rocks in Meso-Neoproterozoic

Foreign studies show that petroleum system in Meso-Neoproterozoic is controlled by climatic conditions and tectonics-paleogeography environment. Green-house period is the main period of hydrocarbon source rocks in Meso-Neoproterozoic since the melting of glaciers can lead to rising sea level, which is beneficial to the accumulation of organic sediments. Many tectonic events in Proterozoic show that there were many continents in that period, and the distance between the continents was much closer than that of the present. As a result, there must be many generalities, similarities and comparabilities among the continents in the sediment characteristics, scale of sedimentary facies, lateral stability of sedimentary sequence, geological processes of tectonic events and the generating petroleum geologic conditions (Wenzhi et al. 2018). There also developed large-scale high-quality hydrocarbon kitchens in North China, Yangtze and Tarim craton (Fig. 2.11) in Meso-Neoproterozoic.

A large number of outcrops and drilling data reveal that the hydrocarbon source rocks with thick thickness and high organic matter abundance developed in the Meso-Neoproterozoic in china, and the maturity is generally high nowadays. The value of Ro distributes from 1.6 to 3.8%, which is in the main gas window of liquid hydro-carbon cracking. High-quality source rocks are found in the outcrops and key wells, which are Chuanglinggou Formation, Hongshuizhuang Formation and Xiamaling Formation in Changcheng System, Madian Formation in Cambrain in North China craton, Datangpo Formaton in Nanhua System and Doushantuo Formation in Sinian in Yangtze Craton, Sinian in Tarim craton (Table 2.2).

Source rocks of Changcheng and Jixian System in Mesoproterozoic are the oldest source rocks found in China, which are mainly found in North China Craton. Through the research of source rocks of outcrops and some drillings in Ordos basin, the result shows that the source rocks of the Changcheng System in the northern margin (Shujigou Formation) have high abundance of organic matter, with an average TOC of 3.8%. The thickness of source rocks is up to 100–400 m, and relatively mature, with

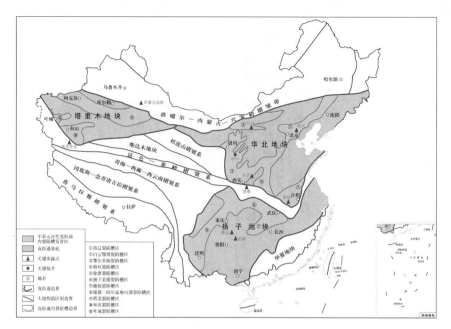

Fig. 2.11 Palaeo-structure and source rocks in Meso-Neoproterozoic in North China, Yangtze and Tarim

the Ro value of about 2.0–3.0%. The average TOC in the Cuizhuang Formation of Changcheng System in the southern margin is 0.52%, with relatively low abundance. The thickness of source rock is 20–40 m and relatively thin. T_{max} averages 580 °C, and equivalent Ro is about 2.5–3.0%. Gray-black mudstone was drilled at 4630–4632 m and 4656–4657 m in Well Tao59, with a cumulative thickness of about 3 m (no penetration). TOC of cuttings pyrolysis analysis is about 3–5%, T_{max} is about 460–500 °C, and the equivalent Ro value is about 1.8–2.2%. From the seismic profile, the source rocks are likely to develop on a large scale (Fig. 2.12).

The source rocks in Datangpo Formation of Nanhua System, Doushantuo and Dengying Formation of Sinian are mainly distributed in Yangtzi Craton (Jian 2005, 2006). The source rocks of calcareous shale in Sinian interglacial period of Hefei Basin in south North China craton have been found recently in this research, with the total thickness over 60 m and three sets of rocks from the bottom to up. The black calcareous shale and moraine rock are inter-developed (Fig. 2.13). The TOC of source rocks is 1.09–3.56%, with the average value of 2.2%. The average value of T_{max} is 508 °C, and the equivalent Ro value is about 2.5%. This set Fm in source rock is in the same period with the mudstone in Luonan Fm in Shanxi and Huangqikou Fm in Ningxia. It may be developed in the southern and western margins of North China Craton from regional correlation, which can be concluded as a set of important source rocks.

Table 2.2 Effective Source Rocks of Meso-Neoproterozoic in three major Cratons in China

Area/Basin		Strata series		Thickness (m)	TOC (%)	R_o (%)	Data location
North China	Hefei	Sanian	Fengtai Formation	>60	1.09–3.56/2.2	2.1–3.7/2.5	Huoqiu, Anhui
			Xiamaling Formation	>260	3–21/5.2	0.6–1.4/1.1	Xiahuayuan, Hebei
	Yanliao	Mesoproterozoic	Hongshuizhuang Formation	>90	1–6/4.1	0.8–2.1/1.6	Kuancheng, Hebei
			Chuanlinggou Formation	>240	0.6–15/2	1.2–2.5/2.2	
	Ordos	Mesoproterozoic	Cuizhuang Formation	>40	0.2–1.5/0.52	2.5–3.0/2.6	Yongji, Shanxi
			Shujigou Formation	100–300	0.8–17/3.8	2.0–3.0/2.2	Guyang, inner Mongolia
			Doushantuo Formtaion	20–40	0.5–14/2.9	2.1–3.8/2.8	Liujing, Zunyi
Yangtze	Neo-Proterozoic		Datangpo Formation	25–35	0.9–6.8/4.4	2.1–2.4/2.3	Songlin, Guizhou
Tarim	Huanan-Sinian			130–320	0.6–4.9/2.9	1.1–1.4/1.2	Kuruktag

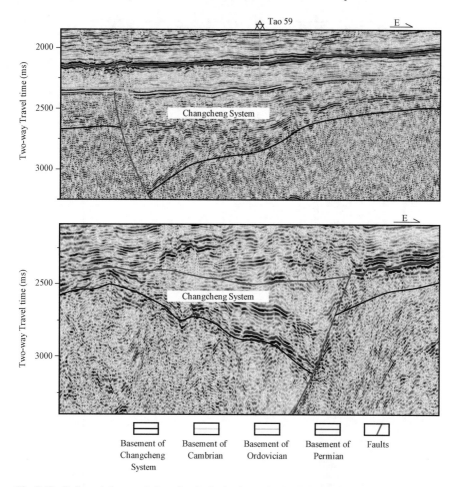

Fig. 2.12 Reflected characteristics of typical seismic section in Changcheng System through Well Tao59 in Ordos Basin

The maturity of Meso-Neoproterozoic source rocks is generally high in the three major cratons in China (Table 2.2), with the Ro value of generally above 2.0%, which has reached a high-over mature stage. According to the viewpoint of "successive gas generation" in organic matter, the Ro value of early kerogen stage is less than 1.6%, and is 1.6–3.2% in the late stage of liquid hydrocarbon cracking. The amount of liquid hydrocarbon cracking into gas is 2–4 times higher than that of the same amount of kerogen. Based on this judgment, the source rocks of Meso-Neoproterozoic are still in the peak stage of liquid hydrocarbon cracking, and the opportunity of oil exploration is relatively small, but the potential of gas exploration is great.

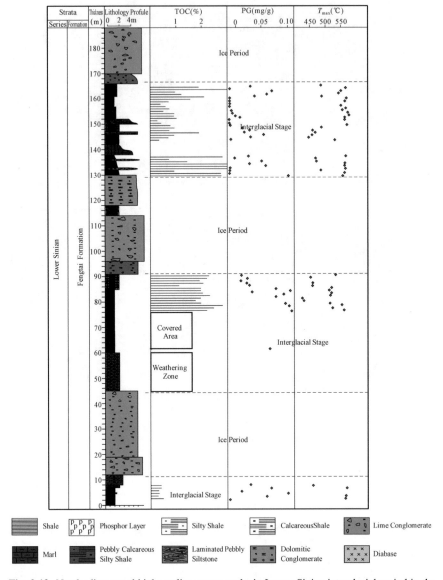

Fig. 2.13 Newly discovered high-quality source rocks in Lower Sinian interglacial period in the south North China Basin

2.2.2 Hydrocarbon Generation Mechanism of Meso-Neoproterozoic Source Rocks

(1) Low micropaleontology flourishment

The earth biosphere was dominated by prokaryote such as archaea and cyanobacteria, and eukaryote such as acritarchs and green algae from Proterozoic to Early Paleozoic, especially before the Cambrian Explosion. Such lower organisms are extremely difficult to preserve in aerobic environment, but can be enriched and stacked in anaerobic environment. The Proterozoic organisms have been very prosperous though they are relatively low organisms, and eukaryotes and prokaryotes have occupied the stage of life. Abundant microorganisms have laid a good material foundation for the enrichment of organic matters and the development of high-quality source rocks (Visser 1991; Tieguan and Keyou 2011; Jian et al. 2012; Zecheng et al. 2014a, b).

① Species of organisms under aerobic and anoxic conditions

There have been two events of oxygen content increase in Proterozoic (2.4-2.2Ga ago) and Neoproterozoic (1Ga ago), which have played an important role in promoting the evolution from prokaryotes to eukaryotes and single-celled organisms to multicellular organisms. However, due to the overall low oxygen content in the atmosphere, the biological populations are dominated by lower organisms such as cyanobacteria, algae and acritarchs.

A material foundation for the enrichment of organic matters has been laid by a variety of microorganism types in Proterozoic in North China, such as prokaryotes, eukaryotes and different forms of acritarchs. Abundant biomakers and many different forms of acritarch fossil assemblages are found in the samples, which indicate the similar bio-combination characteristics of each stage in Proterozoic. Among them, abundant steroids (Fig. 2.14) and spherical algae (Plate I f–i) are found in Chuanlinggou Formation before 1800 Ma indicating the emergence of eukaryotes. Tricyclic terpane (anaerobic bacteria and unknown algae), hopane compounds (prokaryotes or eubacteria membrane lipids), gammacerane (prokaryotes) and a large number of quasi-Kunbu diaphragms similar to brown algae are found in Xiamaling Formation (1 Ga ago, Plate I j–m) and Hongzhuang Formation (Plate I n–o) in Xiahuayuan profile, indicating the extensive reproduction of benthic algae.

② Interglacial Period and Microbial Flourish

There are many glacial-interglacial cycles in the Proterozoic, including Huronian glaciation in Palaeoproterozoic and "Snowball Events" in Neoproterozoic, which are characterized by the alternation of Greenhouse-Ice Chamber environment. The deep-sea organic matter reservoirs formed during the glacial period are released in the interglacial period, which leads to the increasement of water nutrients and the flourish of organisms. There are two reasons for the flourish of organisms in interglacial period: (1) the light carbon released by deep sea organic matters enter

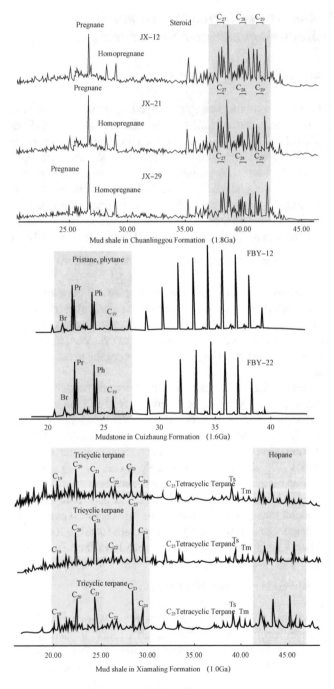

Fig. 2.14 Sterane and Hopane Biomarkers in Meso-Neoproterozoic

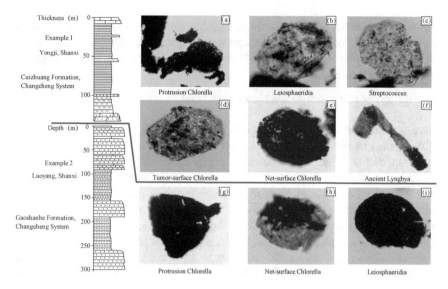

Fig. 2.15   Acritarch fossil data of Cuizhuang Formation (**a–f**) and Gaoshanhe Formation (**g–i**) in the southwest margin of Ordos Basin

the surface ocean by rising ocean current, and then increase the initial productivity of organic matter by organisms reabsorption. (2) large amounts of CO_2 released by deep sea organic matters cause greenhouse effect and melting of glaciers, which lead to the increase of land-surface runoff, nutrients into the ocean and organisms flourish further. Not only a large number of filamentous algae fossils (prokaryotes) from tubular sheath of cyanobacteria, but also a large number of sterane markers and spore microfossils (eukaryotes) larger than 10 μm in diameter are detected in the source rocks of Chuanlinggou Formation in Jixian section and Cuizhuang Formation in Yongji section before 1.6Ga (Fig. 2.15).

The weathering and hydrothermal activities caused by warm climate bring a lot of nutrients to these microorganisms. Warm weather conditions are also linked to the Greenhouse effect caused by CO_2 released by volcanic activities. The research shows that the climatic conditions in North China went through many stages of evolution throughout the whole Proterozoic. Due to the strengthening of evaporation, the climate in Changcheng System of Palaeoproterozoic was transformed from hot-humid climate to hot-dry climate. The climate condition in early-middle Changzhougou Formation to Chuanlinggou Formation was hot-humid climate. By the late Chuanlinggou Formation, the water salinization occurred by the decrease of precipitation and evaporation of water. The salt pseudocrystals related to hot-dry climate didn't appear until the late Tuanshanzi Formation. The basin maintained a high salinity environment by the large evaporation caused by the rising temperature because the paleo-latitudes of North China Plate were closer to the equator in the Gaoyuzhuang Formation. Jixian System of North China Plate was still in the hot-humid climate. However, in 720–635 Ma of Neoproterozoic, the Snowball Earth occurred because the continents located near low latitudes and the whole oceans in the equator were covered by ice through a rapid transition between Glacial Period and Super-Greenhouse Climate. The Snowball Earth was then quickly disappeared under the Super-Greenhouse Climate.

③   Terrigenous denudation and bioconcentration

The paleo-oceanic sediments emerged from the ocean and formed continental crust by the rising crustal movement, and transported back into the ocean again by weathering denudation. When the rift basin caused by the crustal descent was affected by transgression, the surface of the paleo-weathered crust would be deposited again, then formed the unconformities. These unconformities combined the early sediments, and even the denudation products of ancient bedrock from paleo-continent, with the relatively younger oceanic sediments, which may have a time difference of billions of years. Such formation deletions were accompanied with strong weathering denudation, and the surface was eroded by wind and water flow. Chemical weathering released metal ions from the surface into the ocean, making the chemical conditions of the deep ancient ocean change dramatically (Fig. 2.16), and changes in the chemical composition of ocean could be the driving force of life's explosion.

④   Volcanic activities and bioconcentration

Volcanic activities can meet the basic conditions of high-quality source rocks, such as the enrichment and good preservation of organic matters. The formation of high-quality source rocks is effected by biological blooming or death caused by inorganic

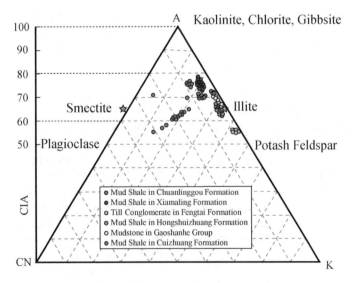

Fig. 2.16 CIA weathering of Proterozoic source rock samples in North China shown in Al_2O_3–$(CaO + Na_2O)$–K_2O (A-CN-K) triangular diagram

salt provided by the addition and hydrolysis of many inorganic elements, which come from volcanic materials erupted by terrestrial or underwater eruptions and descend into water such as lakes and oceans. Landing volcanic ash and the gases contained can protect organic matters by dissolving in the water and forming a reduced environment. Volcanic activities can form a special environment for living organisms, and provide additional generative kerogen rocks (Yan et al. 2016). Nitrogen and phosphorus nutrients, and radioactivities from volcanic activities contribute to microbial prosperity. At the same time, volcanic activities may cause mass biological deaths by releasing harmful substances such as copper and zinc, as well as toxic gases such as hydrogen chloride and chlorine. The soluble gases released by volcanic activities, such as H_2S, CH_4, SO_2 and CO_2, can react chemically with oxygen and water, resulting in ocean stratification by anoxia and gravity differentiation. The underwater environment of hypoxia or reduction is conducive to organic matter preservation as the hypoxia water is distributed in the lower part.

Microbial culture experiments (Fig. 2.17 and Plate II) show that Nitrogen, Phosphorus, Uranium and other nutrients can promote microbial growth and oil accumulation. Generally speaking, with the increase of Phosphorus (K_2HP_4) concentration, the microbial growth rate increases, while with the increase of Nitrogen ($NaNO_3$) and Uranium (U_3O_8) concentration, the microbial growth rate increases at first and then

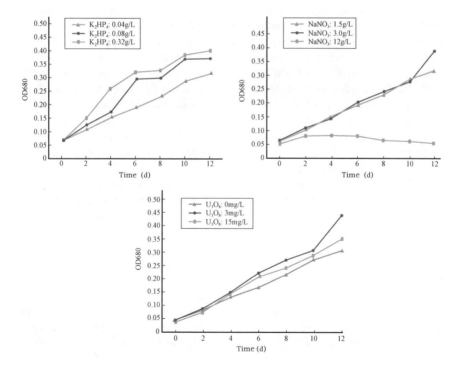

Fig. 2.17 Experiments of microorganism cultivation in culture mediums with different concentration

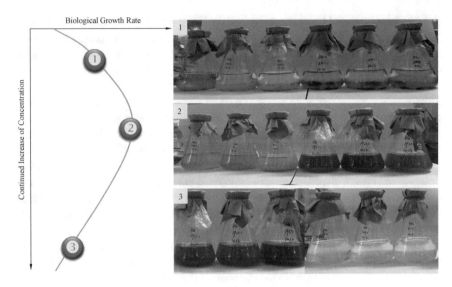

Fig. 2.18 Effects of increased nutrients (N, U) on microorganism growth

decreases (Fig. 2.18). Taking the Nitrogen and Uranium elements as example, the result shows that when the concentration of $NaNO_3$ is 3 and 6 g/L in the first 4 days of the culture period, it has a slight effect on the growth of microcystis, while shows inhibition from the 4th to 9th day. After that, there is an obvious promoting effect. When the concentration of $NaNO_3$ is 12 g/L, the growth of microcystisis significantly inhibited in the whole cultivation process. When the concentration of KNO_3 is 1.5 g/L, it can promote the growth of the porphyridium. And there is no significant difference between the test and control group after 30 days of treatment. However, when the concentration of KNO_3 is 3 and 6 g/L, the growth of the porphyridium is inhibited. Similarly, different concentrations of uranium have the same characteristics for the growth of porphyridium.

High concentration of nitrate has an effect on oil content of microcystis. It directly leads to the death of microcystis when the concentration of Nitrate is 12 g/L, so it is impossible to determine the oil content. But compared with the control group (1.5 g/L), the oil content slightly increases when KNO_3 concentration is 6 g/L. The effect of different nitrate on the oil content of Porphyridium cruentum is that the increase of KNO_3 concentration is beneficial to the accumulation of oil, and the highest oil content can reach 47.17% when the concentration of KNO_3 is 3g/L. With the increase of nitrate concentration, the oil content decreases, but it's still higher than that of the control group. TEM results also show that compared with the normal control group, when the concentration of KNO_3 is 3, 6 and 12 g/L, there is no significant difference in the internal structure of microcystis cells; similarly, the high concentration of KNO_3 does not cause damage to the cells. However, its extracellular sheath layer gradually thickens with the increase of nitrate concentration.

(2)   Organic enrichment mode

There are two kinds of water environments in Proterozoic in North China, and different microbial communities are distributed in different water environments, which also controlled two kinds of enrichment modes of organic matter (Baomin et al. 2007). One is in poor oxygen-anoxic environment, represented by Xiamaling Formation (Fig. 2.19), taking prokaryotes as parent source; and the other is in anoxic-sulfide environment, represented by Hongshuizhuang Formation (Fig. 2.20), which has both prokaryotes and eukaryotes in major source.

① ˙ The organic matter enrichment mode in weak-retention anoxic-sulfied environment

Microbial communities are particularly sensitive to water salinity, high or low salinity is conducive to their survival. Element geochemistry analysis shows that the sedimentary water environment of Hongzhuang Formation is an open-weak retention environment. When the basin transgression occurs, the environment is beneficial to maintain communication between the basin water and the external open water, and the marine water at this time has normal salinity conditions. The water environment with weak retention and suitable salinity is beneficial to the prosperity of eukaryotic phytoplankton flourish. These eukaryotic phytoplanktons provide a large amount of

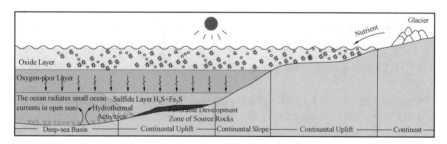

Fig. 2.19 The organic matter enrichment mode in weak-retention anoxic-sulfied environment in Meso-Neoproterozoic

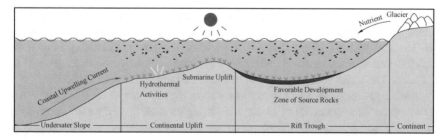

Fig. 2.20 The enrichment mode of organic matters in poor oxygen -anoxic salty environment with strong-retention of Meso-Neoproterozoic

material basis for the formation of organic matters in Hongzhuang Formation, with the spherical algae constituting the main part of these eukaryotes, followed by a small number of special-shaped microbial communities, which not only have many species, but also have a high degree of prosperity. Prokaryote benthos communities are mainly distributed in shallow water environment. The previous biomarkers and fossil data also show that there are both prokaryote benthic algae and eukaryote phytoplankton algae in the sedimentary period of Hongshuizhuang Formation.

②   The organic matter enrichment mode in strong-retention and poor oxygen-anoxic salty environment.

In the water environment with strong retention, the connectivity between the water body of basin and the external open water is weakened, and the recharge of the water body is limited, which leads to the evaporation of the water body larger than the recharge of fresh water, and the salinity of the water body begins to increase, which is not conducive to the growth of planktonic microorganisms. However, this water environment is beneficial to the growth of benthic microorganisms, which can represent the distribution of marine microbial communities in Xiamaling Formation. Biomarkers and fossil data also show that the palaeo-marine organisms in Xiamaling Formation are mainly prokaryotic benthic algae. A large number of nutrients and nutrient salt which are bought by hydrothermal activity, volcanic activity, weathering and upwelling lead to a large explosion of benthic organisms. In the same way, the organic matter of these microorganisms consume most of the oxygen in the water during the sinking process, forming the anoxic environment and leading to the preservation and enrichment of large amount of organic matter.

(3)   Hydrocarbon generation potential of source rocks

Proterozoic source rocks have great potential of hydrocarbon generation, and different types of microorganisms have different influence on the hydrocarbon generation potential of Proterozoic organic matter. The simulation experiments of hydrocarbon generation in gold tube at high temperature and high pressure are carried out on kerogen-like materials of prokaryotes and eukaryotes as well as kerogen of Xiamaling Formation with low maturity. These experiments prove that kerogen-like materials in both prokaryotes and eukaryotes have strong hydrocarbon generation potential, but the hydrocarbon generation potential of eukaryotes is obviously higher than that of prokaryotes, and eukaryotes mainly generate gas, while prokaryotes mainly generate oil. The highest hydrocarbon generation yield of kerogen in Xiamaling Formation with low maturity can reach 301.66 mg/g, which shows high hydrocarbon generation potential. The major source of organic matters in Xiamaling Formation is dominated by benthic prokaryotes, and thermal simulation experiments show that the hydrocarbon generation potential of eukaryotes is higher than that of prokaryotes. It is speculated that the organic-rich source rocks with eukaryotes as the major source in Proterozoic of North China may have higher hydrocarbon generation potential.

The yields of simulation experiences in gaseous, non-gaseous and liquid hydrocarbons from the kerogen-like materials of prokaryotes (microcystis) and eukaryotes (porphyridium) are shown in Fig. 2.21. The kerogen-like materials of two algae-have have both similarity and uniqueness in hydrocarbon potential. The simulated temperature and reaction time of the selected experiments are exactly the same, all representing the hydrocarbon generation characteristics of the low-maturity, maturity and high-maturity stage. Therefore, the simulation results are comparable.

Total hydrocarbon yield: the variation curves of total hydrocarbon yield in the two kinds of kerogen-like materials are basically the same. In the whole process, the total hydrocarbon yield of kerogen in porphyridium has been higher than that in microcystis. The total hydrocarbon yield of kerogen-like materials in the two algae reach the highest at 500 °C, with microcystis 216.22 mg/g, and porphyridium 326.34 mg/g.

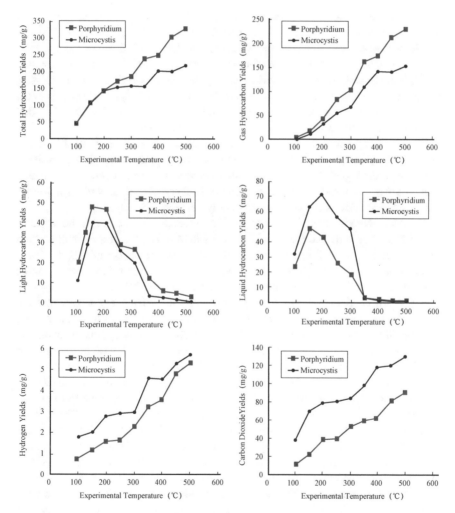

Fig. 2.21 Yield rates of hydrocarbon and non-hydrocarbon of microcystisand porphyridium kerogen-like materials

The results show that both the two algae have strong hydrocarbon generation potential, and the hydrocarbon generation potential of eukaryotes is higher than that of prokaryotes. The yields of gaseous hydrocarbons and liquid hydrocarbons increase with the increase of temperature. At 200–400 °C, the yield of gaseous hydrocarbons increases obviously, and the yield of liquid hydrocarbons decreases sharply, which is due to the C–C bond fracture of liquid hydrocarbons with high molecular weight, resulting in the formation of gaseous hydrocarbons with low molecular weight. The gaseous hydrocarbon yield of porphyridium is always higher than that of microcystis, but the yield of liquid hydrocarbon is lower than that of microcystis. It is speculated that the eukaryotic plankton represented by porphyridium is dominated by gas generation, and the prokaryotic benthos represented by microcystis is dominated by oil generation. This conclusion supports the previous view that "planktonic algaes generate oil and benthic algaes generate gas." Light hydrocarbon yield: the light hydrocarbon peak yields of the two groups of algae reach the maximum at 150–200 °C, which may correspond to the over-maturity stage of wet gas generation. The light hydrocarbon yield decreases sharply, and the pyrolysis occurs, generating a large number of gaseous hydrocarbons over 200 °C. This stage corresponds to the over-maturity stage. The light hydrocarbon yield of porphyridium is higher than that of microcystis in the whole hydrocarbon generation process.

By comparing and analyzing the hydrocarbons generated by kerogen-like materials in porphyridium of eukaryotes and microcystis of prokaryotes, it is considered that both microorganisms have strong hydrocarbon generation potential, but that of eukaryotes is obviously higher than that of prokaryotes. Meanwhile, eukaryotes mainly generate gas while prokaryotes mainly generate oil.

The Meso-Neoproterozoic rift troughs North China craton controll the distribution of Proterozoic source rocks, which shows the broad prospects of hydrocarbon resources. Based on drilling data, field profile and seismic data, combining the distribution characteristics of rift trough, the distribution range and thickness of source rocks in Proterozoic of North China can be predicted (Fig. 2.22).

Some Drilling data of basins and outcrops reveal the existence of source rocks in the development area of rift troughs. However, in areas where drilling data are not available, the distribution of source rocks is constrained mainly by seismic data. The results show that the thickness of Proterozoic source rocks in the Yanliao rift trough is up to 1500 m, and the distribution area is 169×10^4 km^2; the thickness of source rocks in the Dongyu rift trough is up to 1800 m, and the distribution area is 115×10^4 km^2; the thickness of source rocks in Xionger rift trough is up to 1800 m, and the distribution area is 160×10^4 km^2; the thickness of source rocks in Shangan rift trough is up to 2100 m, with the distribution area of 156×10^4 km^2; the thickness of source rocks in Jinshan rift trough is up to 2100 m, with the distribution area of 180×10^4 km^2; the thickness of source rocks in Beiyuan rift trough is up to 900 m, and the distribution area is 148×10^4 km^2. On the whole, the Proterozoic source rocks are developed in six rift troughs of North China, with large thickness and wide distribution area. The hydrocarbon generation potential of Proterozoic hydrocarbon resources in North China is calculated by genetic method. The results show that the overall area of source rocks is 928×10^4 km^2, with the thickest of 2100 m.

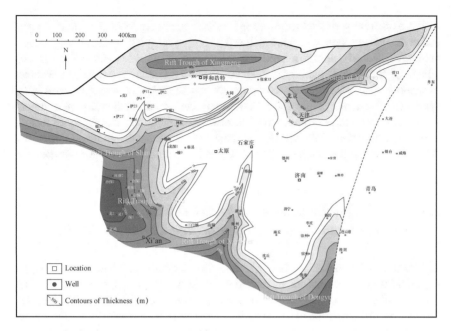

Fig. 2.22   Prediction of Proterozoic source rocks in North China

Among them, the hydrocarbon-generation amount of resources in Yanliao rifting trough reaches 79.4×10^8 t, the hydrocarbon-generation amount of resources in Dongyu rifting trough reaches 80.2×10^8 t, and the hydrocarbon-generation amount of resources in Xionger rifting trough reaches 32.5×10^8 t. Compared with Yanliao rifting trough and Dongyu rifingt trough, the lower hydrocarbon potential of Xionger rifting trough may be due to the material base of organic matters. The organics are deposited in Paleoproterozoic, dominated by prokaryotes with low hydrocarbon potential. And the primary productivity in marine at that time is low. In the evaluation of hydrocarbon resources and the selection of exploration blocks, it is suggested to calculate the hydrocarbon-generation amount of source rocks according to the distribution characteristics of the rifting trough and the types of major kerogen in different strata.

2.3   Source Rocks and Hydrocarbon Generation Mechanism in Salty Environment

There are many examples of symbiosis between evaporite and large-scale high-quality source rock in salty environment, and the corresponding sedimentary models have been established. In the early stage of evaporite formation, the bottom water activities are close to stagnation, as a result, large number of organisms die. With the increase of salinity, the reduction environment is gradually formed, which is beneficial to the preservation of organic matter. The high-quality hydrocarbon source rocks in the continental basins are related to the salty lake basin in Meso-Cenozoic of China (Jinqiang et al. 2008), while the deep small marine cratons are mainly distributed in shallow-water environment. High-quality source rocks are also developed in the symbiotic system of gypsum-salt rocks and carbonate rocks. In the salty environment, the gypsum-salt rocks promote the hydrocarbon generation of kerogen, especially in the high-over maturity stage. At the same time, the magnesium ions in formation fluid can promote the cracking of crude oil and the formation of high sour gas.

2.3.1   High-Quality Source Rocks Are Developed in the Combination of Gypsum and Carbonate Rocks

In addition to the coal deposits in Meso-Cenozoic continental basins of China, the high-quality source rocks are all related to the salty lake basins, such as Bohai Bay Basin, Qaidam Basin, Jianghan Basin, Subei Basin and Pearl River Mouth Basin. The formation and distribution of high-quality source rocks in salty lake basins are studied by sedimentology, geochemical theory and oil source comparison. The results show that the high-quality source rocks in the fractured lake basin and the foreland basin are accompanied by evaporative salt minerals such as carbonate, sulfate and chloride. The favorable environments for organic matter deposition are brackish lakes, lagoons, and salty lakes. The source rocks in Qingshankou Formation Member I and Nenjiang Formation Member I of Songliao Basin are recognized as the main source rocks. Recent studies have shown that they are related to water invasion and may be the product of anoxic environment in the salty lake basin, in which the carbonate content in the organic enriched layer is more than 15%, and the oil shale is composed of clay carbonate lamina and organic-rich clay lamina. The Triassic in Ordos Basin is originally considered as a large freshwater lake deposit, in which the Yanchang Formation consists of five members, and the Member II and III are the concentrated section of source rocks with high content of carbonate, which should be deposited in the salty environment. Under the gravity action, the salinity stratifications are easily formed in the lake. The salinity of surface water is low, which is suitable for the survival of euryhaline and halophilic plankton. The bottom salinity of deep water is high, which is lack of free oxygen and suitable for the preservation of organic matters. The superposition of high yield area dominated by surface water biology and the bottom water hypoxia area is the development area of high-quality source rocks. The sudden flood caused by climate change may destroy the stratified water body, leading to the lake water desalination, and forming the source rocks with low content of organic matters. As a result, the salty high-quality source rocks and the common desalination source rocks are often interbedded, and the more the amount of salty sediment is, the greater the potential of hydrocarbon generation is.

The deep strata in China are mainly formed by shallow sea deposits, and the gypsum-salt rocks in evaporative environment have no effect on hydrocarbon formation. Whether the high-quality source rocks can be developed in the symbiotic system of gypsum-salt rocks and carbonate rocks?

The high-quality source rocks can be developed in swallow water environment by the symbiotic system of gypsum-salt rocks and carbonate rocks, although the evaporites developed in deep-small craton of China are usually formed inside the carbonate platform, with the characteristics of small distribution area, thinner thickness and poor longitudinal continuity. Effective source rocks are developed in the evaporation tidal flat-lagoon environment of Majiagou Formation in Ordos Basin. This research discusses the development characteristics of source rocks in platform gypsum environment through the analysis of sedimentary microfacies and organic carbon based on Well Chengtan1 of Majiagou Formation in Ordovician. This set of source rocks is mainly formed by dark dolomitic mudstone, which belongs to dolomitic mudflat environment. The natural gamma curve is at a relatively high value, around 150 API, which is concentrated in the upper Majiagou Formation Member III, followed by the middle-lower Majiagou Formation Member III and Member I (Fig. 2.23). According to the lithologic data, the monolayer thickness of dolomitic mudstone is generally thin, ranging from 10 to 20 m, and the cumulative thickness is large, accounting for 30–50% of the formation thickness. TOC is generally 0.3–5.14%, with an average of 1.35%. The kerogen carbon isotope is heavy in source rocks, especially in the high-abundance (TOC) source rocks, which is 3–4‰ heavier than that in a low-abundance (TOC) source rocks, reflecting the relationship between salty water environment and salinization degree. Generally speaking, three transgressions and marine regressions are occurred alternately in the Ordovician. The deposit environment of open-shallow carbonate platform is formed during the transgression, but it is not conducive to the formation of source rocks. While the evaporation tidal flat-lagoon environment formed in the regressive period is conducive to the formation of source rocks. The source rocks are distributed around "salt depression" on the plane (Fig. 2.24), which is controlled by the gypsum-bearing dolomitize flat and secondary depression. The effective source rocks can be formed and protected in the small-craton carbonate platform gypsum-salt environment, and the dolomitic mudflat and lagoon are favorable environments for the development of source rocks.

It is also worth studying whether the source rocks under the salty strata of Cambrian in the East Sichuan Basin are developed. The salt-bearing strata in this area are developed in Middle-Lower Cambrian and the source rocks of Qiongzhusi Formation developed under the salty strata (Fig. 2.25). It is worth studying whether the source rocks developed in Canglangpu Formation, that is, before the salinization of Middle-Lower Cambrian. The preliminary judgment shows that there should have the environment condition of source rock development, and if so, the generation potential will be greatly enhanced.

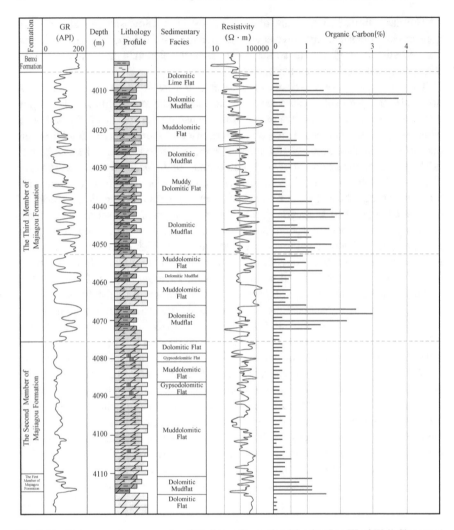

Fig. 2.23 Histogram of source rocks of Majiagou Formation Member I to III of Well Chengtan1 in Ordos Basin

2.3.2   Hydrocarbon Generating Mechanism in Saline Environment

The gypsum-saline environment can promote hydrocarbon generation of source rocks. As a result, the generation potential of organics in gypsum and carbonate rocks combination can be larger. In order to reveal the effect of gypsum rocks on organic hydrocarbon generation, two kinds of simulation experiments are designed, one is the hydrocarbon generation by direct matching of gypsum salt rock and kerogen, and the other is adding different medium conditions to simulate gas cracking of crude oil.

Fig. 2.24 The distribution of effective source rocks developed in the evaporation tidal flat-lagoon environment in Ordos Basin. **a** the thickness map of the effective source rocks in the upper Majiagou Formation Member V; **b** the thickness map of the effective source rocks in the 5–10 sub-sections of Majiagou Formation Member V; **c** the thickness map of the effective source rocks in Majiagou Formation Member III; **d** the thickness map of the effective source rocks Majiagou Formation Member I; **e** Lithofacies palaeogeography of 4 sub-section in Majiagou Formation Member V; **f** Lithofacies palaeogeography of 6 sub-section in Majiagou Formation Member V; **g** Lithofacies palaeogeography of Majiagou Formation Member III; **h** Lithofacies palaeogeography of Majiagou Formation Member I

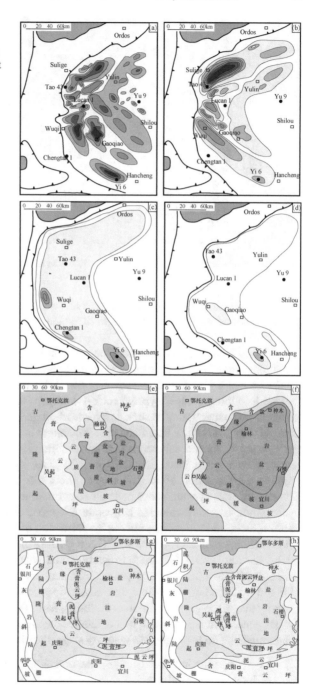

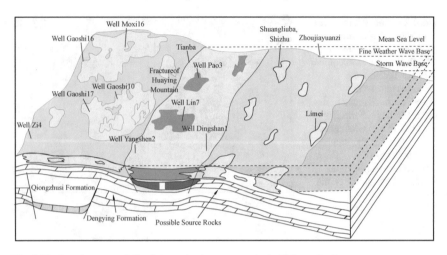

Fig. 2.25 Development of Cambrian subsalt source rocks in Sichuan Basin

(1) Gypsum-salt rocks can promote the hydrocarbon generation of kerogen, which
      is obviously in the high-over maturity stage

The lithology model by simulation experiment is shown in Fig. 2.26. Experimental
conditions are as follows: 35 Mpa of constant pressure, 325 °C of initial temperature,
2 °C/h of heating rate, and measured at one point per 25 °C. The experiment is
divided into three groups: gypsum rock + kerogen, salt rock + kerogen and pure
kerogen. The experimental results of hydrocarbon generation by gypsum salt rock-
kerogen (Fig. 2.27) show that when the temperature is less than 525 °C (Ro <2.2%),
the gypsum rocks and salt rocks inhibit the formation of methane, and when the
temperature is more than 525 °C (Ro >2.2%), the gypsum rocks and salt rocks
promote the formation of methane; when the temperature was less than 475 °C
(Ro <1.8%), the gypsum rocks and salt rocks inhibit the formation of C_{2+} heavy
hydrocarbon gas, and when the temperature is more than 475 °C (Ro >1.8%), the
gypsum rocks and salt rocks promote the formation of C_{2+} heavy hydrocarbon gas;
when the temperature is less than 500 °C (Ro <2.0%), the gypsum rocks and salt rocks
inhibit the formation of liquid hydrocarbon, and when the temperature was more
than 500 °C (Ro <2.2%), the gypsum rocks and salt rocks promote the formation
of liquid hydrocarbon. Therefore, two conclusions can be drawn: in the middle-low
maturity stage, the gas generation is inhibited, and in the high-over maturity stage,
the gas generation is promoted; in the middle-low maturity stage, the oil generation
is inhibited, and in the high-over maturity stage, the oil generation is promoted.

(2) Mg^{2+} in formation fluid catalyzes the cracking of crude oil and the formation
      of high-sour natural gas

The lithology model by experimental simulation is shown in Fig. 2.28. The experimental conditions: constant pressure is 50 Mpa, initial temperature is 365 °C and

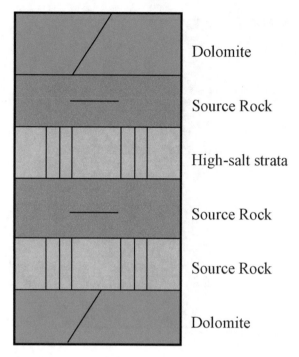

Fig. 2.26 Simulation experience of hydrocarbon generation by gypsum-salt rock and kerogen

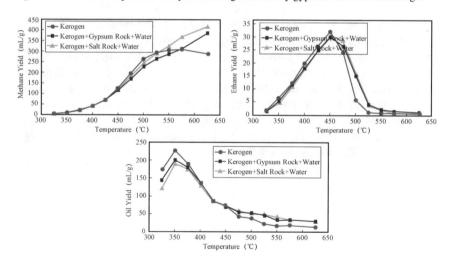

Fig. 2.27 Variation curves of hydrocarbon yield between gypsum-salt rock and kerogen

Fig. 2.28 Gas generation oil-cracking experiences under different medium conditions

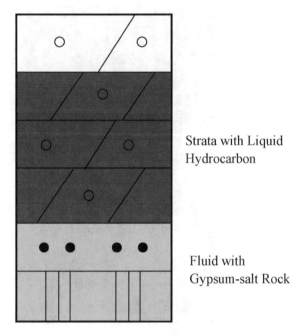

Strata with Liquid Hydrocarbon

Fluid with Gypsum-salt Rock

time range of 0–168 h. The experiment is divided into four groups: crude oil with magnesium sulfate, crude oil with calcium carbonate, crude oil with calcium sulfate and pure crude oil. The model is shown in Fig. 2.28. The experimental results of oil cracking under medium conditions (Fig. 2.29) show that the yields of methane, ethane, propane, and total hydrocarbons are higher in the $MgSO_4$ system than in other systems, and the gas hydrocarbon yields are significantly increased. High-sour gas are formed by TSR in the $MgSO_4$ system, which is similar to the gypsum-salt rock-dolomite composition system with stratigraphic fluid. This system changes the traditional understanding of high sour gas formed by the combination of "gypsum rock + water + hydrocarbon", which reveals the important catalytic effect of Mg^{2+} on the reaction. The process is as follows.

$$C_nH_{2n+2} + SO_4^{2-} + H_2O \ C_nH_{2n+2} + SO_4^{2-} + H_2O \xrightarrow{Mg^{2+}} H_2S \uparrow + CH_4 \uparrow + CO_2 \uparrow$$

2.4 New Generation Model of Coal-Formed Gas

70% of China's proven gas reserves and 63% of its output come from coal strata. However, it has been controversial what the real gas generation potential of coal bearing strata is, and whether the Ro which represents the maturity limit of the end of coal-formed gas generation should be 2.0%, 2.5%, or even higher. In recent

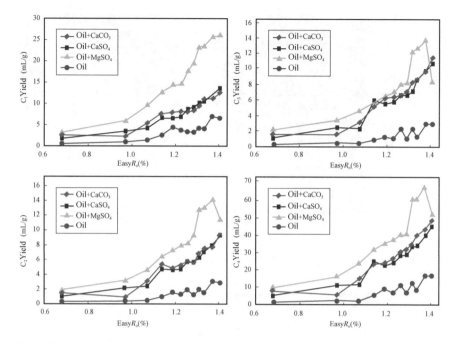

Fig. 2.29 Variation curve of gas generation rate by oil cracking under fluid condition of gypsum salt rock

years, by the hydrocarbon exploration in the upper Paleozoic of the Ordos Basin, Mesozoic of Kuqa Basin and Xujiaweizi regions of the Songliao Basin, the new understanding of hydrocarbon has been expanded and a complete maturity sequence is formed (Fig. 2.30). The mechanism and potential of coal-formed gas generation are re-understood through the "double increase" hydrocarbon generation model of coal in geological conditions, which is established by a series of high-temperature pyrolysis simulation experiments, product analysis and balanced elemental material calculation.

The gas generation of coal-formed source rocks is mainly affected by the maturity of organic matters and the composition of macerals. At present, the method used to measure the gas generation of source rocks is mainly the thermal simulation experiment. Shuichang et al. (2013) have determined the generation amount of coal-formed gas by prolonging the heating time and measuring the H/C ratio of coal in different maturity series. The hydrocarbon gases of coal in the Jurassic of Ordos Basin (Ro is 0.52%) are simulated by two different heating methods (programmed heating method and step-by-step constant-temperature heating method) using gold tube system (Fig. 2.31). The maximum total gas generation of coal measured by the former is less than 200 m^3/t, and it measured by the latter is 330 m^3/t. According to the relationship between H/C ratio and maturity in coal proposed by Chen et al., the theoretical maximum gas generation of coal with maturity of 0.5% when Ro reaches

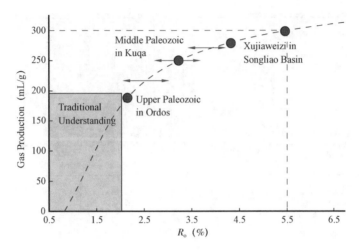

Fig. 2.30 Complete maturity sequence of coal-formed gas in China

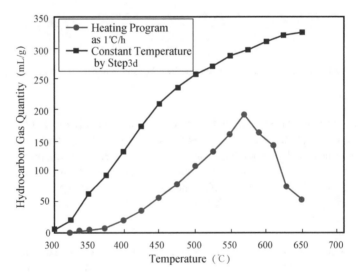

Fig. 2.31 Relation between hydrocarbon gas quantity and simulated temperature of low-maturity coal under different experimental methods

6.0% is 383 m³/t, which is higher than that simulated by experiments. The result is of great significance to the evaluation of coal-formed gas generation and resources in coal-bearing basins in China.

As "all-weather" gas source rocks, coal and coal series mudstone make the formation of coal-formed gas runs through the whole coal-forming evolution. However, there are still some disputes about the gas production rate of coal and coal series mudstone in each evolution stage (Yan et al. 2012). In the early stage, it is generally believed that the Ro value of the "gas death line" in coal is close to 2.5%. Foreign scholars believe that the organic matters of gas source rocks in coal can be recombined to form new kerogen in the high evolution stage, and a large amount of natural gas can be generated when the Ro reaches 5% in the higher evolution stage. Jianping et al. (2007, 2008) believe that the hydrocarbon generation rate of highly enriched III type organic matters represented by coal is low, and the mature stage of hydrocarbon generation continuation is long, with no obvious gas peak, and the lower limit Ro of gas maturity in humus coal can reach 10%. Zhang Shuichang et al. simulated the gas generation of seven coal samples from Songliao, Qinshui and Ordos basins with maturity Ro value from 0.56 to 5.32% by the gold tube thermal simulation experience (Fig. 2.32). The experimental results show that coal can still generate gas before the Ro value is 5.32%, and it is more suitable to set the maturity Ro limit value at 5.0%.

This research establishes the "double increase" hydrocarbon generation model of coal under geological conditions (Fig. 2.33) by deducing the simulation experimental results to the geological conditions combined with the research of gas generation potential, mechanism, time limit, the calibration method and hydrocarbon generation kinetics of simulation maturity experiments.

Compared with the previous gas generating amount of coal and the end of hydrocarbon generation, the maturity limit of gas generation in coal is delayed by 1 time, and the amount of gas is increased by 40–50%.

The first stage is the interaction stage between organisms and thermal effects, that is, the immature evolution stage with Ro less than 0.5%. Bio-fermentation by organics and biogas reduced by CO_2 in the anaerobic environment are produced

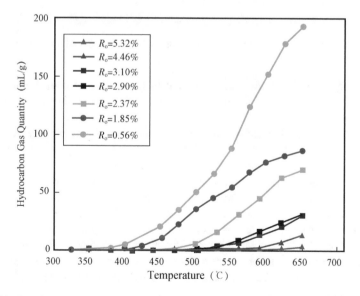

Fig. 2.32 Relationship between hydrocarbon gas quantity generated by coal and simulated temperature at different levels of maturity

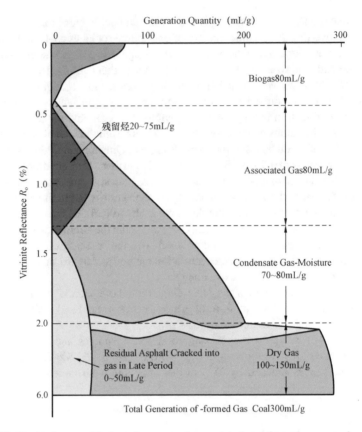

Fig. 2.33 Double-increased hydrocarbon generation model of coal-formed source rocks

in this stage, probably with a small amount of gas generated with thermal effect. According to foreign data, the natural gas production in the biochemical stage is 48–85 m³/t.

The second stage is the traditional main oil-window stage, namely the mature stage with Ro of 0.5–1.3%. In this stage, the long-chain alkyl, aliphatic ring and some aromatic groups with low bonding force in coal structure fall off and form natural gas under thermal action. In addition to methane, there are heavy hydrocarbon gases such as ethane and propane, as well as a small number of liquid hydrocarbons in the natural gas. The amount of gas generation at this stage is about 80 m³/t.

The third stage is the main gas generation stage, namely the high to over-mature stage with Ro of 1.3–2.5%. In this stage, the short chain alkyl in the coal fractures and detaches to form gas, or the long-chain hydrocarbons which have been formed but have not been barred are cracked to form gas. The amout of generation gas in this stage is 80–120 m³/t.

The fourth stage isw the evolution stage of over-maturity dry gas, with Ro greater than 2.5%. A significant reduction process in carbon isotopes of methane at this stage indicates that the fracture of methyl from aromatic methyl or alkyl hydroxybenzene in macromolecular structures is the mechanism for methane generation. The process gradually becomes dominant as the degree of thermal evolution increases. As the main component of lignin and cellulose, alkyl hydroxybenzene is abundant in the vitrinite of coal. As a result, a considerable amount of natural gas can still be generated at high maturation stage. The gas volume generated at the stage is 100–150 m^3/t, accounting for 30–50% of the total gas generating amount.

The formation of coal-formed gas is a continuous process. It can generate gas between organic depositions to graphitization, but the amount, kerogen, gas composition and mechanism of gas generation are different in various stages. Because of the particularity of kerogen in coal source rocks, it contains a large number of alkyl phenols compounds and materials with aromatic structure. Only in high-temperature environment can these substances be cracked. The branched chain breaks from the aromatic structure to form hydrocarbon gas, and the polycondensation of aromatic structure is further developed. Therefore, phenolic compounds in the over-maturity stage are the main kerogen. They can still produce a large amount of natural gas, which has important application value for the resource evaluation of coal-formed gas in high-over-mature coal-bearing basins.

References

Baomin, Zhang, Zhang Shuichang, Bian Lizeng, et al. 2007. Brief analysis on the development model of marine source rocks of Neoproterozoic-Lower Paleozoic in China. *Chinese Science Bulletin* 52 (Supplement 1): 58–69.

Chen, J.P., H.P. Ge, X.D. Chen, et al. 2008. Classification and origin of natural gases from Lishui Sag, the East China Sea Basin. *Science China: Earth Sciences* 51 (Supplement): 122–130.

Digang, Liang, Guo Tonglou, and Chen Jianping. 2008. Some progresses on studies of hydrocarbon generation and accumulation in marine sedimentary regions, southern China (Part 1): Regional distribution of four sets of marine source rocks, South China. *Marine Origin Petroleum Geology* 13 (2): 1–16.

Jian, Wang. 2005. New Advances in the study of "the Nanhua system"—with Particular reference to the stratigraphic division and correlation of the Nanhua system, South China. *Geological Bulletin of China* 24 (6): 491–495.

Jianping, Chen, Zhao Wenzhi, Wang Zhaoming, et al. 2007. A discussion on the upper limit of maturity for gas generation by marine kerogens and the utmost of gas generative potential_Taking the study on the Tarim Basin as an example. *Chinese Science Bulletin* 52 (A01): 95–100.

Jian, Zhang, Shen Ping, Yang Wei, et al. 2012. New understandings of Pre-Sinian sedimentary rocks in the Sichuan Basin and the significance of oil and gas exploration there. *Natural Gas Industry* 32 (7): 1–5.

Qiang, Jin, Zhu Guangyou, and Wang Juan. 2008. Deposition and distribution of high-potential source rocks in seline lacustrine environments. *Journal of China University of Petroleum* 32 (4): 19–23.

Shuichang, Zhang, Hu Guoyi, Mi Jingkui, et al. 2013. Tine-limit and yield of natural gas generation from different origins and their effects on forecast of deep oil and gas resources. *Acta Petrolei Sinica* 34 (Supplement 1): 41–50.

Tieguan, Wang, and Han Keyou. 2011. On Meso-Neoproterozoic primary petroleum resources. *Acta Petrolei Sinica* 32 (1): 1–7.

Visser, J. 1991. Biochemical and molecular approaches in understanding carbohydrate metabolism in Aspergillus Niger. *Journal of Chemical Technology & Biotechnology* 50 (1): 111–113.

Wenzhi, Zhao, Wang Zhaoyun, Zhang Shuichang, et al. 2005. Successive generation of natural gas from organic materials and its significance in future exploration. *Petroleum Exploration and Development* 32 (2): 1–7.

Wenzhi, Zhao, Zhang Guangya, and Wang Hongjun. 2005. New achievements of petroleum geology theory and its significances on expanding oil and gas exploration fields. *Acta Petrolei Sinica* 26 (1): 1–7.

Wenzhi, Zhao, Wang Zecheng, and Wang Yigang. 2006. Formation mechanism of highly effective gas pools in the Feixianguan Formation in the NE Sichuan Basin. *Geological Review* 52 (5): 708–718.

Wenzhi, Zhao, Wang Zhaoyun, Wang Hongjun, et al. 2006. Cracking conditions of oils existing in different modes of occurrence and forward and backward inference of gas source rock kitchen of oil cracking type. *Geology in China* 33 (5): 952–965.

Wenzhi, Zhao, Wang Zhaoyun, Wang Dongliang, et al. 2015. The status and significance of dispersed liquid hydrocarbon accumulation. *Petroleum Exploration and Development* 42 (4): 401–413.

Wenzhi, Zhao, Hu Suyun, Wang Zecheng, et al. 2018. Petroleum geological conditions and exploration status of Proterozoic-Cambrian in China. *Petroleum Exploration and Development* 45 (1): 1–13.

Yan, Zhao, and Liu Chiyang. 2016. Effects of volcanic activity on the formation and evolution of hydrocarbon source rocks. *Geological Science and Technology Information* 35 (6): 77–82.

Yan, Song, Zhao Mengjun, Hu Guoyi, et al. 2012. Progress and perspective of natural gas grochemistry researches in China. *Bulletin of Mineralogy, Petrology and Geochemistry* 31 (6): 529–542.

Zecheng, Wang, Jiang Hua, Wang Tongshan, et al. 2014. Hydrocarbon systems and exploration potentials of Neoproterozoic in the Upper Yangtze region. *Natural Gas Industry* 34 (4): 27–36.

Zecheng, Wang, Jiang Hua, Wang Tongshan, et al. 2014. Paleo-geomorphology formed during Tongwan tectonization in the Sichuan Basin and its significance for hydrocarbon accumulation. *Petroleum Exploration & Development* 41 (3): 305–312.

Zhao, W.Z., Z.Y. Wang, S.C. Zhang, et al. 2008. Cracking conditions of crude oil under different geological environments. *Science in China Series: Earth Sciences* 51 (Supplement 1): 77–83.

Chapter 3
Formation Mechanism of Deep Reservoir

The large-scale effective reservoir is the critical factor to the large-deep oil and gas fields, with the complex reservoir properties such as the reservoir space type, geometry, mechanism of dfviagenesis and preservation caused by the multi-stages diagenetic reformation and increased heterogeneity of deep reservoir. The large-scale reservoirs can be developed in all the three kinds of deep rock reservoirs, but they have different formation mechanisms and controlling factors. As a result, it is significant to summarize the main controlling factors of the formation of different deep reservoirs for the research on accumulation condition and hydrocarbon distribution, as well as the selection of favorable exploration areas of deep oil and gas.

3.1 Accumulation Mechanism of Deep Carbonate Reservoir

The deep-layer pores in carbonate reservoirs can still be formed and maintained due to the unnecessary correspondence between the reservoir properties and buried depth, as the carbonate reservoirs is not limited by the depth, which is different from the clastic reservoirs. The formation of deep pores is controlled by dominant sedimentary facies. And the large deep oil and gas fields with large-scale reservoirs can be formed on large-area high energy facies belts by superposition of the constructive diagenesis such as late-stage dissolution and burial. The oil and gas reserves in the combination of gypsum-salt rocks and carbonate rocks are huge. The combinations of gypsum-salt rocks and carbonate rocks are distributed in three major marine craton basins of China, and three types of combination are developed. Reservoirs with different scales can be developed in different combinations of gypsum-salt rocks and carbonate rocks, whether in a superficial or buried environment. It has been proved that the microbial carbonate rocks in Proterozoic-Cambrian strata have good reservoir properties in Sichuan, North China, Tarim and other areas of China, due

© Petroleum Industry Press 2021
S. Hu and T. Wang, *Deep-Buried Large Hydrocarbon Fields Onshore China: Formation and Distribution*, https://doi.org/10.1007/978-981-16-2285-4_3

to preservation of preexisting pores by the early-stage dolomitization and microbial outgassing.

3.1.1 Dominant Sedimentary Facies in the Late Superposition of Constructive Diagenesis

(1) Carbonate rocks developed in large-area layered reef-beach facies are the base of large-scale reservoirs

With the development of long-term and complex diagenetic superposition and transformation, the original sedimentary materials of deep carbonate rocks are very important to the reservoir evolution. The reservoirs of reef-beach facies are the major part of deep carbonate reservoirs in both limestone karst reservoirs and dolomite pore-cave reservoirs (Guangyou et al. 2006; Baomin and Jingjiang 2009; Ling et al. 2013). Reef-beach facies of the high-energy facies belts, especially beach facies, developed in domestic sedimentary basins, are the material basis of the formation of deep carbonate reservoirs (Yigang et al. 2004; He and Luo 2010; Wei et al. 2012). The multi-stage karst reconstruction in high-energy facies belts is the main controlling factor for the formation of large-scale carbonate reservoirs. There are two types of high-energy facies belts developed in carbonate strata, such as reef facies and beach facies, while the scale of beach facies is larger than that of reef facies. Two types of high-energy facies belts are developed in 12 sets of strata in the three major basins, such as the reef-beach facies on the platform margin with the distribution area of $(13–16) \times 10^4$ km^2, and the beach facies in the platform with the distribution area of 23.30×10^4 km^2.

The reef-beach complex in platform margin belt is developed in the high-energy environment around the rift in the craton basin. The reef and grain banks in the platform margin belt of Proterozoic-Cambrian strata are superposed developed, with a thickness of 20–150 m, a width of 4–20 km, and a length of hundreds to thousands of meters, such as the platform margin belt of Dengying Formation in Sichuan Basin.

The large-scale grain beachs are developed in the high-energy environment around the evaporation lagoon. In the gentle carbonate slope, the grain beaches are symmetrically distributed on both sides of the lagoon, and stacked with a thickness of 30–120 m and an area of $(1–8) \times 10^4$ km^2. Such as the grain beaches of Longwangmiao Formation in Sichuan Basin, with an area of 8×10^4 km^2, and the large-distributed grain beaches of Lower Cambrian in Tarim Basin, with an area of platform-margin reef of 8123 km^2, and an area of reef and beach within platform of 87,557 km^2.

Three stages of beaches are vertically developed in the gentle-slope platforms of Longwangmiao Formation of Sichuan Basin, which are located in the platform uplift area on both sides of the platform depression or lagoon on the plane, with the characteristics of double beaches. The cumulative thickness of the beaches is 30–90 m, and the thickness of the grain beach reservoir is 10–70 m. The grain-beach reservoirs of Longwangmiao Formation formed in the syngenetic metasomatism are mainly composed of sandy dolomite and a small amount of oolitic dolomite, with the original rocks of bioclastic sandy limestone and oolitic limestone. The reservoir spaces are mainly composed of intergranular pores, intracrystalline (dissolution) pores and dissolution pores and caves, with little fractures. The maximum porosity is 11.28%, mainly distributed between 2 and 8%, the maximum permeability is 101.80 mD, and mainly distributed between 0.01–10 mD. The distribution area of grain beaches in Longwangmiao Formation is nearly 3×10^4 km^2, and the reserves in Gaoshiti-Moxi area is more than trillion cubic meters. The geological background of gentle slope and development model of double beaches are developed in Xiaorbulak Formation of Lower Cambrian in Tarim Basin similar to Longwangmiao Formation of Lower Cambrian in Sichuan Basin, with the distribution along the edge of lagoon and depression, and the area of beaches within platform of 9×10^4 km^2 (Table 3.1).

(2)   The constructive transformation of reservoir is the key factor for large-scale reservoir in late period

The large-scale effective carbonate reservoirs must be reformed by late-stage diagenesis. While the formation of high-quality carbonate reservoirs in deep strata are mainly controlled by epigenetic karstification and, erosive fluid dissolution of various causes, TSR, burial dolomitization, structural hydrothermal dolomitization and so on during the burial process.

① Large area of quasi-layer karst (weathering crust) reservoirs

Controlled by the unconformity of large-scale buried hill, interbedded karst interface and fracture system, the quasi-layer karst (weathering crust) reservoirs are widely distributed in Yijianfang Formation and Yingshan Formation of south margin in Tabei (Hu Mingyi et al. 2014), Yingshan Formation in Tazhong (Fig. 3.1), the Upper combination of Majiagou Formation in Ordos Basin, top of Maokou Formation and Leikoupo Formation in Sichuan Basin, including two types of karst reservoirs, namely carbonate buried hill and inner reservoir.

The Yingshan Formation limestone of Ordovician, on low uplift in Lunnan of Tabei area, covered by Carboniferous sandstone and mudstone, is the product of regional tectonic movement, with the characteristics of peak and hill, and the buried hill peak of hundreds of meters, which represents the strata denudation and loss up of 120 Ma. The karst fractures and caves under the unconformity are important reservoir spaces for oil and gas. They are distributed in the range of 0–100 m below the unconformity and formed in the period of supergene karstification, with a distribution area of 2×10^4 km^2 and a reserves scale of several hundred million tons.

The composite dolomites of Cambrian-Penglaiba Formation in Yaha-Yingmaili area of Tabei, Leikoupo Formation in Sichuan Basin and Upper Majiagou Formation of Ordos Basin are respectively covered by clastic rocks of Kapshaliang Group of Jurassic, Xujiahe Formation of Triassic and Benxi Formation of Carboniferous, representing long-period strata denudation and missing. The geomorphologic fluctuation

Table 3.1 Statistics table of reef and beach facies in three major basins

Basin	Strata	Reef and beach in platform margin		Beach within platform		Exploration findings
		Distribution	Area (10^4 km^2)	Distribution	Area (10^4 km^2)	
Tarim Basin	Lianglitag Formation	Tazhong, Bachu, Tabei,Tangnan	1.2	Tazhong, Bachu	0.3	Tazhong
	Yijianfang Formation	Tazhong, Bachu, Tabei,	1–1.5	Tabei, Tazhong, Gucheng	1.5	Halahatang
	Yingshan Formation	Lunnan, Tazhong, Gucheng	2–3	Whole Basin	3–5	Halahatang, Tazhong, Gucheng
	Penglaiba Formation	Lunnan, Gucheng	2–3	Whole Basin	3–5	Well Tazhong 162, Well Gucheng 4
	Cambrian	Lunnan, Gucheng	2–3	Generally Developed		Well Zhongshen 1
Sichuan Basin	Jialingjiang Formation			Whole Basin	5–6	More than 40 small gas fields
	Feixianguan Formation	North Sichuan	0.25–0.35	Central Sichuan, North Sichuan, East Sichuan	2.5–3	Longgang, Luojiazhai, Tieshanpo
	Changxing Formation	North Sichuan	0.15–0.25	Central Sichuan, North Sichuan, East Sichuan	0.8–1.2	Longgang, Huanglongchang, Wubaiti
	Qixia-Maokou Formation	Northeast Sichuan	0.6	Central Sichuan, West Sichuan	0.8–1.5	Well Long 16, Well He 3, Well Shuangtan 1
	Cambrian			Central Sichuan, East Sichuan	3–5	Longwangmiao, Anyue
	Sinian	Central and East of Sichuan	2.6	Central Sichuan	0.6–0.8	Central Sichuan
Ordos Basin	Ordovician	Southwest Margin	0.2–0.5	East—Central Part	3–5	Well Chun 2, Well Xuntan 1

is minimal; the peak and hill are obscure, and the fracture-cave system is undeveloped, which may be related to the fact that the dolomite in the surface freshwater environment is more difficult to dissolve than the limestone. The dolomite reservoirs space are mainly composed of intercrystalline pores, intercrystalline dissolution pores, intergranular pores, algal-framework pores and anhydrite-moldic, which

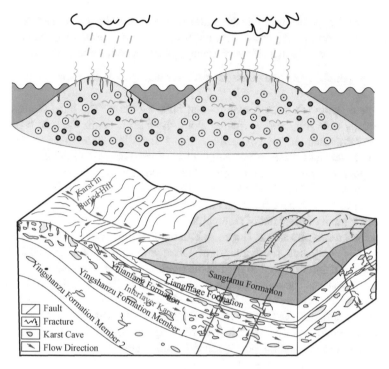

Fig. 3.1   Sketch map of development of syngenesis karst reservoir and interlayer karst reservoir

are the product of superimposed reconstruction of the existing dolomite reservoir. The distribution area of dolomite reservoirs with weathering crust in Yaha-Yingmaili area, north Tarim Basin, is 200 km^2, the oil-bearing area is 36 km^2, and the reserves is 2000×10^4 t; the distribution area of dolomite weathering crust reservoir in Jingbian area of Ordos Basin is 2×10^4 km^2, and the reserves of natural gas is up to trillion cubic meters. In fact, there are two possible origins for the formation of dolomite reservoir space under weathering crust. First, the dolomite reservoirs are formed before the formation of weathering crust, that is to say, they have already developed as porous dolomite reservoirs before they are lifted to the surface. Second, the dolomite reservoirs are formed during the period of weathering crust karstification, but the characters of dolomite strata lifted to the surface are closely related to the transformation effect of karstification in weathering crust. The more soluble lime the dolomite strata contain, the more karst pores are formed, and the better transformation effect is. The size and distribution of dissolution pores are determined by the occurrence and distribution of limestone matter, and the preservation of dissolution pores are supported by the insoluble dolomite. Therefore, the origin of the reservoir space of dolomite weathering crust is significant to reservoir prediction. For example, the reservoirs are independent to the weathering crust if the major part of reservoirs are formed before the weathering crust. The reservoirs are mainly distributed below

0–100 m of the unconformity if the major part of reservoirs are formed during the karst period of weathering crust. The quality of dolomite reservoirs developed under the unconformity is the best if the reservoir space is a composite of them.

② Them thick layers and palisade of Karst fracture-cave reservoirs

This type of karst reservoir is mainly controlled by faults. The karst fractures and caves of the reservoirs are distributed along the fault zone, in the type of net and palisade, rather than the type of para-laminar. The vertical distribution span of fractures and caves in the reservoirs are much larger as no obvious sequence missing or stratigraphic unconformity are developed in the continuous stratigraphic sequence. The reservoirs in Yijianfang Formation and Yingshan Formation of Halahatang area, North Tarim, Yijianfang Formation and Yingshan Formation of Yingmai 1–2 well block are typical of this kind of reservoirs.

The karst reservoirs are developed in Yijianfang Formation and Yingshan Formation in Halahatang area, south margin of Tabei, with an area of nearly 10,000 km², and reservoir depth of over 200 m. The distribution of reservoirs space is controlled by fracture system. The development of caves is controlled by main faults, and the development of pores and caves is controlled by fracture system. The farther away from the main faults, the less developed the pores and caves (Fig. 3.2). Karst fractures and caves are mainly formed in the strike-slip faults, the associated with fracture system and the karstification period related to the faults.

The sedimentary sequences of Yijianfang Formation and Yingshan Formation in Yingmai Block 1–2 of northern Tarim are complete, and covered by Tumuxiuke Formation, Lianglitag Formation and Sangtamu Formation, with no obvious stratigraphic missing and unconformity. The Yingmai No. 2 structure is developed with domal structure, with a structural area of 7 km² and a structural amplitude of 560 m. Three groups of faults are developed, one is NNE large scale strike-slip fault, with long extension, cutting the middle-upper Cambrian-Silurian strata. The other two groups are small-scale faults with NNW and NWW trending, mainly developed in the height of the dome and cutting the Ordovician strata, with a distribution area of 63 km². Dissolution pores and caves are developed mainly along the faults and fractures in the high part of the dome, with no development of matrix pores of the surrounding rocks.

③ Quasi-layer—palisade dolomite reservoirs

The quasi-layer—palisade dolomite reservoirs are developed in Qixia-Maokou Formation in Sichuan Basin, with the characteristics of stratified distribution. The original rocks of dolomite reservoirs are reef-beach facies rocks. However, not all the reef-beach facies rocks have been dolomitized into reservoirs, and both underground and outcrop data reveal that reef-beach reservoirs without dolomitization are tight and nonporous. The dolomitization ratio is related to the development degree, initial

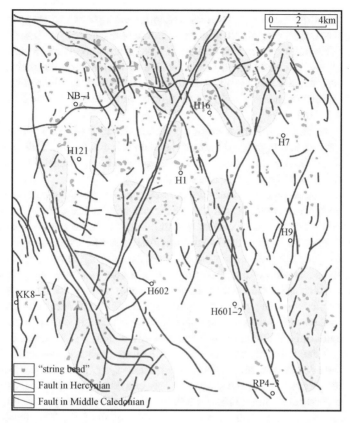

Fig. 3.2 Distribution relationship between karst "string bead" and faults of Paleozoic strata in Halahatang area, north Tarim Basin

porosity and buried-hydrothermal activities of reef-beach facies, and the dolomitization ratio of Member 2 of Qixia Formation and Member 2 and 3 of Maokou Formation is 20–25%.

The research on the origin of dolomite reservoirs in Qixia-Maokou Formation reveals that, the porous reef-beach facies deposition (mainly located under the sequence boundary or interlayer exposed surface) is the material basis for the development of dolomite reservoir, and fracture system and interlayer exposed surface are passages for the burial-hydrothermal fluid, which results in the dolomitization reef-beach bodies are distributed along the fractures and interlayer exposed surface in a quasi-layer-palisade pattern. And dolomitization does not occur in the tight reef beach far away from the fracture system and the interlayer exposed surface. The dolomitization reef-beach bodies are mainly composed of crystalline dolomite (fine, medium and coarse crystal), with high degree of automorphism and almost no retain original rock structure. The reservoirs developed with pores and caves and reservoir space are mainly composed of intercrystalline pores, intercrystalline dissolution pores and dissolution pores and caves. The dissolution pores are developed with diameter of 1–8 cm, and related to the erosion of buried hydrothermal fluid. The reservoirs are usually filled with hydrothermal mineral of saddle dolomites, with the single-layer thickness of 3–4 m, the cumulative thickness of 20–30 m, and the porosity of 5–10%.

The Lower Yingshan Formation in the eastern Tarim Basin also belongs to this kind of reservoir, and high-yield industrial gas flow is obtained in Well Gucheng 6, Gucheng 8 and Gucheng 9. This type of reservoir can be distributed in a large area, with a poor horizontal continuity and strong heterogeneity.

In a word, three types of reservoirs are developed in the deep carbonate rocks, including reef-beach reservoir, karst reservoir and dolomite reservoir, with obvious characteristics of facies control. Reef (mound)-beach facies is the material basis of reservoir development. The reef-beach (dolomite) reservoirs are widely distributed with quasi-layer pattern in the margin of rimmed platform, the margin of rift inter-platform, the depression of the platform uplift area along gentle-slope platform and the lagoon, which are related to the sedimentary background of rimmed platform and gentle-slope platform. The large-scale quasi-layer karst (weathering crust) reservoirs is related to the large paleouplift and unconformable tectonic setting, which can be distributed not only in the buried hill and peri-clinal area of paleouplift, but also in the interlayer karst area of carbonate strata. The karst fracture-cave reservoirs with thick-palisade distribution are related to the fracture system. The further away from the fracture system, the less the pores and caves develop, and the greater the depth span of the karst reservoirs, as the main fault controls the development of the caves and the fracture system controls the development of the hole. The quasi-layer—palisacle dolomite reservoirs are related to reef-beach facies belt, interlayer exposed surface and fault system.

3.1.2 Reservoir Combination of Gypsum-Salt Rocks and Carbonate Rock

The reservoirs combined with gypsum-salt rocks and carbonate rocks are important fields for oil and gas exploration. There are 115 gypsum-salt sedimentary basins in the world, and 97 basins are related to oil and gas, 66 of which are rich in oil and gas (Schroder et al. 2010). The discovered recoverable oil reserves related to gypsum-salt rocks are 665×10^8 t, and the recoverable natural gas reserves are 103×10^{12} m^3, the reserves of oil and gas in reservoirs combined with gypsum-salt rocks and carbonate rocks are account for 65% of the total reserves of marine carbonate rock. Strata combined with gypsum-salt rocks and carbonate rocks are widely developed in the three major marine basins of China, with a low degree of exploration and recognition, and it will become an important replacement field once a breakthrough is made. Strata combined with gypsum-salt rocks and carbonate rocks are developed in all the three major marine carton basins of China, which are mainly concentrated in three sets of strata, such as O, C-P and T_{1+2}.

(1) Development characteristics of gypsum-salt rocks in two typical areas

① Cambrian gypsum-salt rocks in middle and upper Yangtze Region

A symbiotic system of Cambrian gypsum-salt rocks and carbonate rocks is developed in the middle-upper Yangtze area, composed of carbonate rocks, sulfate rocks, halides and thin-layer different amounts of fine-grained sand and mudstone. Moreover, the gypsum-salt rocks are developed with the characteristics of wide distribution, large thickness variation, thin thickness of single layer, large number of interlayers, more gypsum rocks and less salt rocks.

The formation environment of Cambrian gypsum-salt rocks

The Cambrian gypsum-salt rocks in the middle and upper Yangtze area are mainly developed in the Longwangmiao of Lower Cambrian and Middle Cambrian, and the formation environment of carbonate rocks associated with gypsum-salt rocks is the basis for the development of the gypsum-salt rocks. On the research of field outcrop section as well as the detailed observation and analysis of core, it is considered that the formation environment of carbonate rocks associated with gypsum-salt rocks is tidal flat and shoal, while the sedimentary environment of gypsum-salt rocks is interbank lagoon.

The strata contained gypsum-salt rocks in the middle-upper Yangtze area are mainly composed of Longwangmiao of Lower Cambrian, Middle Cambrian and Upper Cambrian. The rocks of the Longwangmiao Formation in lower Cambrian are composed of medium to thick limestone and argillaceous limestone in light and dark gray, with a small amount of sandstone, shale, dolomite and oolitic limestone. The light gray dolomite with sand debris or oolitic are mianly developed in Longwang-miao Formation in ancient uplift of Leshan-Longnusi. In slope-depression, the lower part is composed of gray limestone while the upper part is composed of gray dolomite interbedded with gypsum-salt rock. In middle part of middle and upper Yangtze plat-form, Middle Cambrian are mainly composed of micrite dolomite and argillaceous dolomite, while the strata in middle and upper Yangtze platform are composed of micrite dolomite. The beddings with vein ripple and lenticular (Fig. 3.3), and the sedi-mentary structures such as laminated structure, ripple mark, pinnate cross-bedding, gypsum and halite pseudocrystal structure are widely developed in these carbonate rocks. It is reflected that the rocks in the sedimentary structures are the products of shallow water environment, and the development environment is the clear water tidal-flat.

According to the analysis of outcrop section and core, the vein, ripple and lentic-ular bedding are common developed in Longwangmiao Formation, Shilongdong Formation and Gaotai Formation, which indicate that tidal-flat environment is devel-oped in Longwangmiao Formation of Early Cambrian and Middle Cambrian in

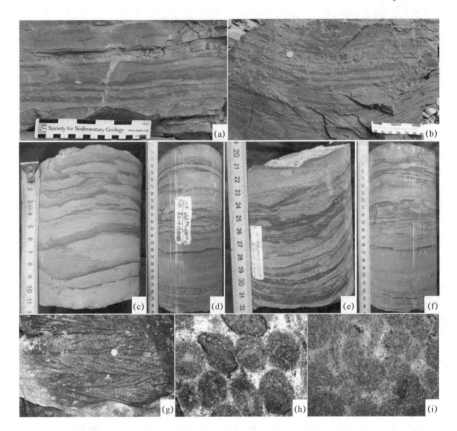

Fig. 3.3 Identification marks of Cambrian tidal-flat environment in the middle and upper Yangtze area **a** flaser bedding, Shilongdong Formation, Shiliucun, Shizhu, Chongqing; **b** flaserbedding, Gaotai Formation, Tuhechang, Xishui, Guizhou; **c** lens, Longwangmiao Formation, Well Moxi 19; **d** lens, Longwangmiao Formation, Well Li 1; **e** lens, Longwangmiao Formation, Well Moxi 39; **f** lens, Longwangmiao Formation, Well Gaoshi 17; **g** pinnate cross-bedding, Yiling, Yichang, Hubei, Tianheban Formation of Lower Cambrian; **h** oolitic dolomite, Longwangmiao Formation, 4528.11 m, Well Gaoshi 17; **i** sandy dolomite, Longwangmiao Formation, 4688.12 m, Well Moxi 19

middle and upper Yangtze area. The sedimentary environment can be well indicated by the morphology of stromatolite. Generally speaking, with the formation environment of weaken hydrodynamic condition, the layered stromatolite is produced in the upper intertidal zone. with the formation environment of strong hydrodynamic condition, the columnar stromatolite is producted in the lower intertidal zone and upper subtidal zone with the reverse flow. Pinnate cross-bedding is one of the typical cross-beddings in tidal environment, developed with straight or slightly upward bendings, and opposite tendency of bendings in the adjacent oblique strata. It can be extended to the interface of the strata, and intersected with each other at acute angles, with the shape of feather or herringbone. Gypsum crystal imprint and halite pseudocrystal are often developed in the Cambrian strata of the middle and upper Yangtze area, which indicates the increase of water salinity during the precipatation of halite crystal in the warm climate. In addition to the tidal-flat environment, the shoal environment are well developed in Cambrian especially in Longwangmiao Formation of Lower Cambrian in middle and upper Yangtze area. The rocks in shoal environment are mainly composed of oolitic limestone and sandy limestone which are often dolomized to form oolitic dolomite and sandy dolomite (Fig. 3.3).

From the Longwangmiao Formation, middle-upper Yangtze area started to enter a stable thermal subsidence stage and ended at Cambrian. During this period, carbonate rocks are deposited with large thickness. And the platform environment in shallow carbonate background was formed, which showes that the Cambrian gypsum-salt rocks in middle-upper Yangtze area are developed as evaporite platform. The development of shoal and lagoon in carbonate platform provides a good environment for the formation of gypsum-salt rocks. Such as the three sets of gypsum-salt deposition developed in Bandenggou Section of Taiyuan, Pengshui, Chongqing (Fig. 3.4). However, no gypsum-salt deposition are found in the adjacent Shiliucun Section of Shizhu, Chongqing, only 3.9 km away from Bandenggou Section in a straight line, while shoal deposits are developed (Fig. 3.5).

Formation conditions of Cambrian gypsum-salt rocks

The necessary hydrological condition for all the evaporating sedimentary environment is the evaporation exceed the recharge amount. There are two basic conditions for the formation of the symbiotic system of gypsum-salt rocks and carbonate rocks: one is the rapid rise and fall of sea level; the other is the high value of Mg/Ca.

The frequent alternation of open shallow sea and limited shallow sea caused by rapid rise and fall of sea level provides favorable conditions for the formation of gypsum-salt rocks. Three third-order regressive sequence cycles with upward shallowing are developed in the symbiotic system of carbonate and gypsum-salt rock of Cambiran in middle-upper Yangtze area, including a third-order layer series of Longwangmiao Formation, consisted of 3–4 secondary cycles upward shallowing.

The gypsum-salt rocks in Longwangmiao Fm of Lower Cambrian in the middle-upper Yangtze area are mainly distributed in the upper part, and associated with dolomite. Dolomite of Longwangmiao Formation is vertically distributed in the middle-upper part, and in contemporaneous-penecontemporaneous diagenetic stage, have undergone the evaporation concentration dolomitization seepage reflux dolomitization and burial dolomitization in shallow burial stage, of which the major one is seepage reflux dolomitization. These dolomites require a hot and dry climate and water environment with a high Mg/Ca.

Sedimentary model of gypsum-salt rocks in Cambrian

The sedimentary model of symbiotic system of gypsum-salt and carbonate rocks and sedimentary model in Longwangmiao Formation of Early Cambrian in middle-upper Yangtze area are developed according to the types and characteristics of sediments, and lithofacies palaeogeographic distribution (Fig. 3.6). The restricted platform, open platform, slope and basin are successively developed from west to east. During the

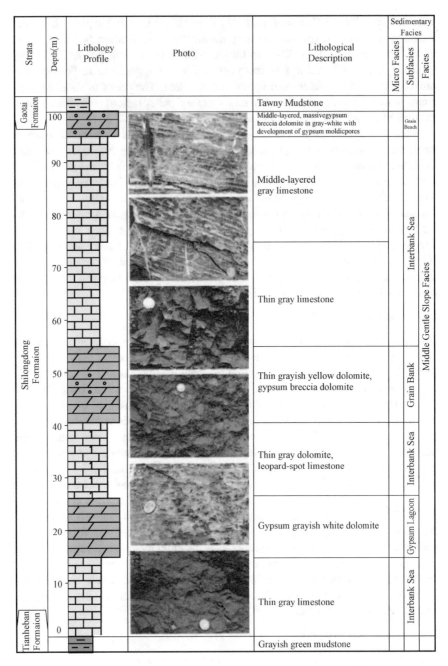

Fig. 3.4 Comprehensive solumnar section of Shilongdong Formation in Lower Cambrian of Bandenggou, Taiyuan, Pengshui, Chongqing

Fig. 3.5 Deposits of oolitic beach in Shiliucun Section, Shizhu, Chongqing

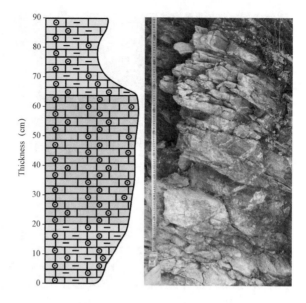

period of relatively high sea level, the shoals in the carbonate platform are developed, including upper shoal and lower shoal. The restricted platform is developed between the shoals, with gypsum-lagoon deposits dominant gypsum-salt rocks accumulated in the low-lying areas. During the period of relatively low sea level, a large area of gypsum lagoons are developed with the barriers formed by shoals. The ancient land, shoreside mainly deposited by quartz sandstone, mixed tidal flat mainly composed of fine-grained carbonate and fine-grained clastic rock, and restricted platform mainly composed of crystal dolomite, grain dolomite and gypsum-salt rock are developed from west to east, from Longwangmiao Formation to late Cambrian in southwest of Sichuan Basin.

② The gypsum-salt rocks of Cambrian strata in Sichuan Basin

In the traditional sedimentary diagenetic model of carbonate gentle-slope grain beach-lagoon in Sichuan Basin, grain beach are developed on the sea side of evaporation lagoon, and the land side is developed in Sabkha environment. The drilling results of Longwangmiao Formation show that the high energy grain beach are developed symmetrically on both sides of the evaporation lagoon, and the grain beach on the land side is more developed. According to the sedimentary phenomena revealed by exploration, the sedimentary diagenetic model of carbonate gentle-slope grain beach-lagoon (gypsum-salt rocks and carbonate rocks) is built up (Fig. 3.7), and the traditional understanding of "carbonate platform paleogeomorphology is flat, sedimentary differentiation is small, and high-energy facies belts are not developed in Sinian-Cambrian carbonate platform of upper Yangtze" is changed. Based on that, the diagenetic model of grain beach reservoir in evaporation environment is

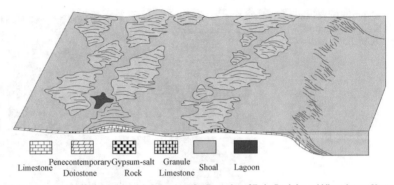

(a) Sedimentary model of high water level in Longwangmiao Formation of Early Cambrian, middle and upper Yangtze area

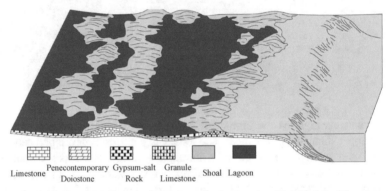

(b) Sedimentary model of low water level in Longwangmiao Formation of Early Cambrian, Yangtze area

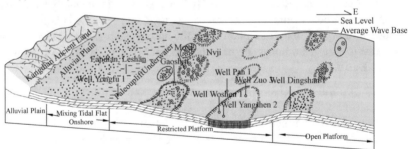

(c) Formation model of gypsum-salt rock from Longwangmiao Formation in Early Cambrian to late Cambrian in the Southwestern Sichuan Basin

Fig. 3.6 Sedimentary model of gypsum-salt rocks in the middle-upper Yangtze area

built up, with the characteristics of large-area grain beach in early highstand system tract, evaporation lagoon and strong dolomitization and dissolution in late highstand system tract to become high-quality reservoir.

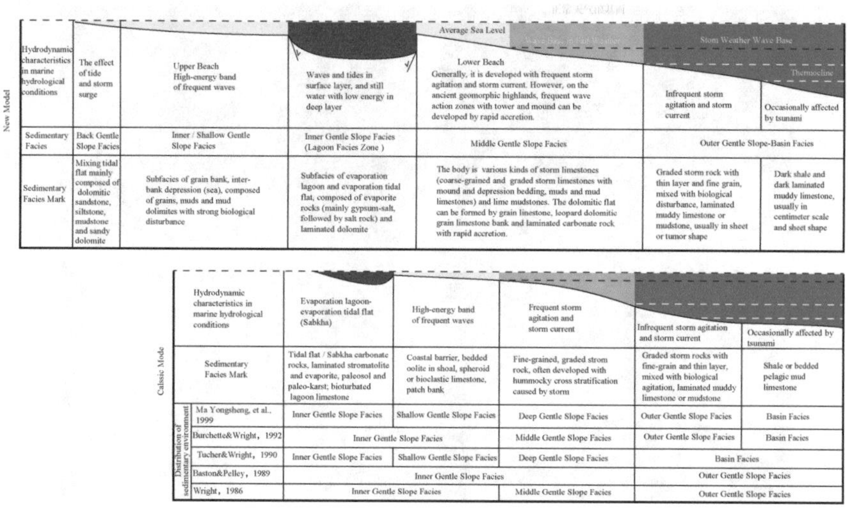

New Model	Hydrodynamic characteristics in marine hydrological conditions	Sedimentary Facies	Sedimentary Facies Mark
The effect of tide and storm surge	Back Gentle Slope Facies	Mixing tidal flat mainly composed of dolomitic sandstone, siltstone, mudstone and sandy dolomite
Upper Beach High-energy band of frequent waves	Inner / Shallow Gentle Slope Facies	Subfacies of grain bank, inter-bank depression (sea), composed of grains, muds and mud dolomites with strong biological disturbance
Waves and tides in surface layer, and still water with low energy in deep layer	Inner Gentle Slope Facies (Lagoon Facies Zone)	Subfacies of evaporation lagoon and evaporation tidal flat, composed of evaporite rocks (mainly gypsum-salt, followed by salt rock) and laminated dolomite
Lower Beach Generally, it is developed with frequent storm agitation and storm current. However, on the ancient geomorphic highlands, frequent wave action zones with tower and mound can be developed by rapid accretion.	Middle Gentle Slope Facies	The body is various kinds of storm limestones (coarse-grained and graded storm limestones with mound and depression bedding, muds and mud limestones) and lime mudstones. The dolomitic flat can be formed by grain limestone, leopard dolomitic grain limestone bank and laminated carbonate rock with rapid accretion.
Infrequent storm agitation and storm current	Outer Gentle Slope-Basin Facies	Graded storm rock with thin layer and fine grain, mixed with biological disturbance, laminated muddy limestone or mudstone, usually in sheet or tumor shape
Occasionally affected by tsunami		Dark shale and dark laminated muddy limestone, usually in centimeter scale and sheet shape

(Banner labels: Average Sea Level — Storm Weather Wave Base — Thermocline)

Classic Mode					
Hydrodynamic characteristics in marine hydrological conditions	Evaporation lagoon-evaporation tidal flat (Sabkha)	High-energy band of frequent waves	Frequent storm agitation and storm current	Infrequent storm agitation and storm current	Occasionally affected by tsunami
Sedimentary Facies Mark	Tidal flat / Sabkha carbonate rocks, laminated stromatolite and evaporite, paleosol and paleo-karst; bioturbated lagoon limestone	Coastal barrier, bedded oolite in shoal, spheroid or bioclastic limestone, patch bank	Fine-grained, graded strom rock, often developed with hummocky cross stratification caused by storm	Graded storm rocks with fine-grain and thin layer, mixed with biological agitation, laminated muddy limestone or mudstone	Shale or bedded pelagic mud limestone

Distribution of sedimentary environment:

Reference					
Ma Yongsheng, et al., 1999	Inner Gentle Slope Facies	Shallow Gentle Slope Facies	Deep Gentle Slope Facies	Outer Gentle Slope Facies	Basin Facies
Burchette&Wright, 1992	Inner Gentle Slope Facies		Middle Gentle Slope Facies	Outer Gentle Slope Facies	Basin Facies
Tucher&Wright, 1990	Inner Gentle Slope Facies	Shallow Gentle Slope Facies	Deep Gentle Slope Facies	Basin Facies	
Baston&Pelley, 1989	Inner Gentle Slope Facies			Outer Gentle Slope Facies	
Wright, 1986	Inner Gentle Slope Facies		Middle Gentle Slope Facies	Outer Gentle Slope Facies	

Fig. 3.7 Comparison the double grain-beach sedimentary models of carbonate gentle slope with the classical sedimentary model in Longwangmiao Formation, Sichuan Basin

The development characteristics of the Cambrian gypsum-salt rocks in the east Sichuan Basin

The gypsum-salt rocks are developed in the regressive sedimentary sequence of middle-lower Cambrian, which represents the sedimentary sequence from shoal to tidal flat and lagoon, the sequence becomes shallower upward. The gypsum-salt strata are composed of silty dolomite, dolomite interbeded with gypsum, and gypsum mixed with halite. The surface of gypsum-salt strata is marked by gypsum breccia, halite pseudocrystal and secondary gypsum. The gypsum-salt rocks in east Sichuan easin are mainly developed in Gaotai Formation, as well as in upper highstand system tract of Longwangmiao Formation in southeast Sichuan, which associated with dolomite.

Field outcrop of gypsum-salt rocks

The gypsum-salt rocks are widely developed in the upper Gaotai Formation in the Xishuihoutan Section, and the associated fold structures are also well developed, which is characterized by the box shape and chevron fold. the gypsum-salt rocks are usually developed as gypsum breccia on the surface, with a gravel diameter of 2 cm. gypsum-salt rocks are often weathered into earthen yellowish clay on the surface (Fig. 3.8). The Lower Shilongdong Formation in Bandenggou Section of Pengshui, Chongqing is covered with vegetation, with the outcrop of 100 m in middle-upper part exposed. Three fourth-order cycles are developed, each of which can be devided into transgressive system tract and highstand system tract. The transgressive system tract of each fourth-order cycle is mainly composed of porphyritic limestone and micrite limestone, and the high level system tracts of each fourth-order cycle is mainly composed of thick dolomite with gypsum-salt rock, with each gypsum-salt

Fig. 3.8 Field photos of gypsum-salt rocks in East Sichuan **a** Gypsum breccia of Gaotai Formation in Xishuihoutan Section; **b** characteristics of transgression-highstand system tract in fourth-order cycle of Shilongdong Formation, Bandenggou Section of Pengshui, Chongqing; **c** gypsum breccia of Shilongdong Formation in Bandenggou Section of Pengshui, Chongqing; **d** thin-layer gray and white gypsum-bearing dolomite, Shilongdong Formation in Bandenggou Section of Pengshui, Chongqing

rock of 10–15 m。The boundry of transgressive system tract and highstand system tract is clearly developed with the interface of thin gray limestone and thin gray and yellow dolomite, while the gypsum-salt rocks on the surface are developed as gypsum breccia dolomites with muddy thin-layer dolomites and gypsum gray and yellow dolomites.

Development characteristics of gypsum-salt rocks in single well

Dolomitic oolitic limestone is developed in the bottom of Shilongdong Formation, lime dolomite and dolomite are developed above, and gypsum-salt rock is developed in the middle-upper part in Well Zuo 3. Sandy dolomite is developed in the bottom of Qinjiamiao Formation, and gypsum-salt rock with great thickness is developed in the middle-upper part in Well Jianshen 1, which reflect the change of sedimentary environment from a relatively open sandy beach to a long-period closed salt lake. The stratigraphic section of Shilengshui Formation in Middle Cambrian of Well Dingshan 1 shows that, the lower section of the gypsum-bearing rock is mainly composed of grey oolitic limestone, microcrystalline dolomite, muddy dolomite and gypsum, which can form a cycle and reflect the salty process of the sea water, as well as represent an upward shallower sedimentary sequence from the shoal to the tidal flat and Sabha on the tide.

Development characteristics of gypsum-salt rock in connecting-well section

The connecting-well section from South Sichuan Basin to East Sichuan Basin reveals that, the gypsum-salt rock of Longwangmiao Formation in Lower Cambrian is mainly developed in South Sichuan Basin and gradually thinned to East Sichuan Basin, such as Well Woshen 1, Well Gongshen 1, Well Lin1, Well Dingshan 1 and Well Zuo 3 et al. Connecting-well section in Sichuan Basin reveals that the gypsum-salt rocks in Middle Cambrian are mainly developed in the east, with characteristics of gradually thickening from south to east, such as Well Ning 2, Well Jianshen 1 and so on (Fig. 3.9).

Detailed horizon calibration of seismic section

The horizons in which the gypsum-salt rocks vertically developed can be defined through the calibration from known wells and the introduction layers in the seismic section, as the response characteristics of plastic aggregation in the gypsum-salt

Fig. 3.9 Comprehensive histogram of gypsum-salt rock in Qinjiamiao Formation of Well Jianshen 1

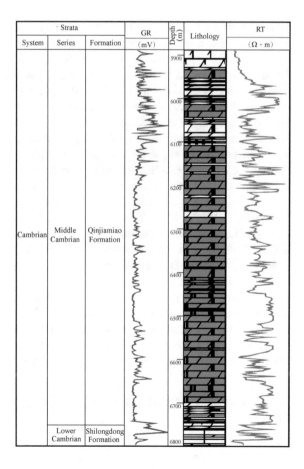

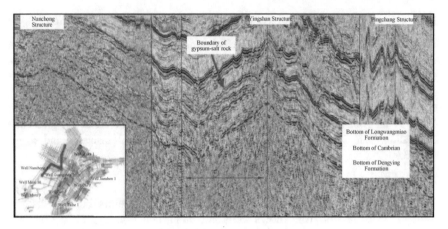

Fig. 3.10 Structural-seismic section of Nanchong-Yingshan-Pingchang

rocks are obvious. From the structural—seismic section of Nanchong-Yingshan-Pingchang, it can be seen that Longgang area is located at the boundary of the development of gypsum-salt rocks. According to the calibration of Well Nanchong 1 and the introduction layers, it can be seen that the gypsum-salt rocks are mainly developed in the Gaotai Formation of Middle Cambrian (Fig. 3.10). The strong trough response characteristics are found in the strata developed with gypsum-salt rocks, but not developed in the strate without gypsum-salt rocks.

Through the NW–SE section of Longgang area, it can be seen that the gypsum-salt rocks are mainly developed in Gaotai Formation through the detailed horizon calibration. The response characteristics of strong trough and plastic aggregation are identified in the seismic sections where the gypsum-salt rocks developed, while they are not identified in the seismic sections without the development of gypsum-salt rocks (Fig. 3.11). The similar characteristics are also revealed in the 3D seismic section of Longgang area. Through the detailed horizon calibration and introduction layers, it can be seen that the gypsum-salt rocks are mainly developed in Gaotai Formation, with the response characteristics of strong trough and plastic aggregation in the seismic section.

Therefore, the gypsum-salt rocks in eastern Sichuan Basin are mainly developed in Gaotai Formation of Middle Cambrian, and only partly developed in southeast Longwangmiao Formation, according to the analysis of detailed horizon calibration and introduction layers in seismic sections through drilled wells in central Sichuan Basin and the analysis of connecting-well sections in south and east part of Sichuan Basin.

Plane distribution characteristics of the Cambrian gypsum-salt rocks in East Sichuan Basin

The plane distribution of the Cambrian gypsum-salt rocks in East Sichuan Basin are mainly revealed by the seismic data. The distribution map of the strata can

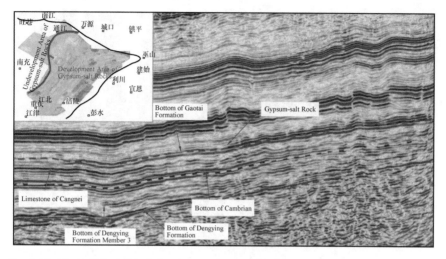

Fig. 3.11 Line LG3d—T900 of NW–SE 3D seismic section in Longgang area

be carried out based on the the thickness of the plastic strata, which is calculated by the regional formation velocity and the top and bottom interface of plastic strata identified by the response characteristics of plastic aggregation showed in the seismic sections (Fig. 3.12). Revealed by drilling and seismic sections, it can be seen that the gypsum-salt strata are developed in the Middle Cambrian of East Sichuan Basin, with the greatly various thickness, which are mainly controlled by two factors of the deposition and late-stage compression. The plane map reflects the thickness change of gypsum-salt bearing stratum, not of pure gypsum-salt stratum. It can be seen that the gypsum-salt rocks of Middle Cambrian are not developed in Central Sichuan Basin, and the development characteristics in East Sichuan Basin are as following: first, the development of the Middle Cambrian gypsum-salt rocks are controlled by Huayingshan fault, as a result, the gypsum-salt rocks are relatively developed in the east of Huayingshan fault while not in the west side. Therefore, the gypsum-salt deposition is obviously controlled by the fault extension. Secondly, the gypsum-salt bearing strata of Middle Cambrianin in East Sichuan Basin is developed regionally. The thickness of the gypsum-salt bearing strata is gradually increased from the middle of Sichuan Basin to the east of Sichuan Basin, and the thick gypsum-salt bearing strata are mainly formed by compaction.

(2)   Combination of gypsum-salt rocks and carbonate rocks

Three types of combination of gypsum-salt rocks and carbonate rocks are developed in small deep-marine craton basins in China, according to the characteristics of lithology change (Table 3.2). Type 1 is composed of carbonate rocks with interlayers of gypsum-salt rocks, with the typical example in Cambrian, Tarim Basin; type 2 is composed of carbonate rocks interbeded with gypsum-salt rocks, with the typical

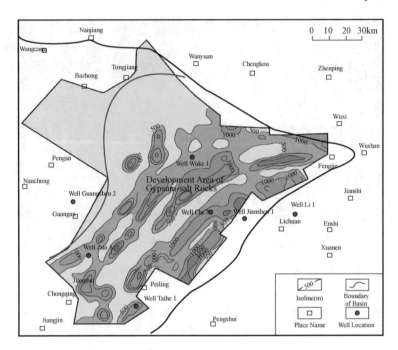

Fig. 3.12 Plane distribution of gypsum-bearing strata in middle-lower Cambrian of East Sichuan Basin

example in Majiagou Formation of Ordovician, Ordos Basin; type 3 is composed of carbonate rocks and gypsum-salt rocks with interlayers of clastic rocks, with the typical example in Longwangmiao Formation of Cambrian, Sichuan Basin.

(3)   Simulation experiment of combination of gypsum-salt rocks and carbonate rocks

① Combination of gypsum-salt rocks and carbonate rocks (combination type 2): simulation experiment of dissolution in supergene environment.

A typical symbiotic system composed of gypsum-salt rocks and carbonate rocks is developed in Majiagou Formation (O_m) of Ordovician, middle-east part of Ordos Basin (Fig. 3.13), in which evaporite rocks are mainly developed in the Member 1, 3 and 5, and carbonate rocks are mainly developed in the Member 2, 4 and 6. The formation and distribution of weathering crust reservoir in the Submember 1–4 of Member 5 (O_m^5) of Majiagou Formation are affected by the supergene dissolution, and developed as the main productive layers of Jingbian Gasfield. The effect of karst on the pore structures of gypsum salt—carbonate reservoirs is quantitatively evaluated based on the dissolution simulation experiment under geological conditions.

Experimental conditions: Screening the samples. The difference of dissolution effects are analyzed by seven groups of physical simulation experiments in supergene

Table 3.2 Comparison of combination of gypsum-salt rocks and carbonate rocks in three major craton basins

Factor	Type 1	Type 2	Type 3
Age	Sinian-Cambrian	Ordovician	Cambrian, Carboniferous
Lithology	Carbonate rocks are developed above and below gypsum-salt rocks	Gypsum-salt rocks are interbeded with carbonate rocks	Terrigenous clastic rocks such as marlstones, mudstones and siltstones are developed above and below gypsum-salt rocks
Distribution	Large area and large thickness	Large thickness and large vertical variation	Large thickness and terrigenous clastic rocks (gypsum mudstones)
Salt Formation Cycles	Simple cycle and less salt-bearing system	Multi-period repetition and many salt-bearing systems	
Tectonic Setting	Craton basins in stable continental plate	Rift-fault basin in the margin of continental plates	
Sedimentary Environment	Restricted basin in epicontinental sea (platform syncline area) or inland shelf sea area	Shallow sea environment of platform margin (marginal depression area)	Bay or shallow shelf adjacent ancient land
Lithologic Association	Carbonate rocks-gypsum-salt rocks-carbonate rocks	Carbonate rocks-gypsum carbonate rocks-gypsum carbonate rocks	Carbonate rocks-terrigenous clastic rocks-gypsum-salt rocks and terrigenous clastic rocks-carbonate rocks
Examples	Cambrian in Tarim Basin	Majiagou formation in Ordos Basin	Longwangmiao Formation in Cambrian, Sichuan Basin and lower Carboniferous in Tarim Basin
Legend	Lithologic association	Lithologic association	Lithologic association

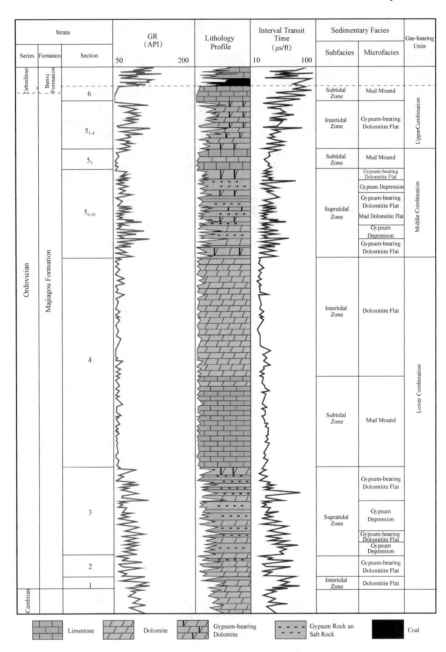

Fig. 3.13 Stratigraphic column of O_{1m} in the middle-east Ordos Basin

karstification, with the salty bearing dolomite, gypsum-bearing dolomite, muddy bearing dolomitic gypsum, dolomitic gypsum and dolomite bearing gypsum in O_m of East Ordos Basin, selected by the content of gypsum-salt rocks. Among them, the gypsum-salt particles are mostly filled in the pores of the salty bearing dolomite and the gypsum bearing dolomite. The porosity of the reservoir is mainly 1–2%, with the anhydrite distributed in mudd-bearing dolomitic gypsum, dolomitic gypsum and dolomite bearing gypsum as nodules. Cylindrical samples with a diameter of 1 cm are prepared by selecting different lithologic samples based on the content difference of dolomite, halite, calcite, gypsum and other major minerals. The reservoir characteristics before dissolution are quantitatively tested by multiple methods. The pores and microfractures with a diameter of centimeter to micro-nano scale are developed in the reservoirs of gypsum-salt rocks and carbonate rocks. The distribution characteristics of pore diameter of the samples before dissolution are quantitatively evaluated mainly based on the methods of N_2 adsorption, CO_2 adsorption and quantitative analysis of mercury injection, combined with SEM observation. The temperature of dissolution condition and the solution properties are the major factors affecting the closure of carbonate reservoirs. Geological history temperature can be determined by the different geographical location of the research area. This simulation experiment mainly recovers the content of CO_2 in the surface leaching fluid of the research area by the content of atmospheric CO_2 in geological history period, which can represent the property of solution fluid, according to the strong dissolution effect of CO_2 on carbonate reservoirs. The fluid before karstification shows weak acidity, with the pH 6.0–6.4. The test conditions of supergene karst are generally determined as atmospheric pressure, saturated CO_2 water (6–15%) and air temperature of 30 °C. The karst experiment conditions in the closed environment are determined by the surface temperature and fluid properties of the research area. The containers are selected by the number of samples, and connected in series, then the samples of 50 g are put into the sealers respectively, adding deionized water of 500 mL in order to saturate the CO_2. Then the excess CO_2 is fed to the liquid in the next sealer by inserting a catheter into the sealer. According to the method above, the supersaturation of CO_2 in the solution can be reached in each container, showing the property of weak acidity. Then the karst experiment can be run by puting the container into the oven and setting the temperature and time. Multiple methods are used to quantitatively measure the characteristics of reservoirs after dissolution.The quantitative test of the dissolved reservoirs composed by gypsum-salt rocks and carbonate rocks is carried out by the means of N_2 adsorption, CO_2 adsorption and mercury injection, and the distribution and connectivity of pores in dissolved reservoirs are quantitatively evaluated with the combination of SEM and CT analysis. Quantitative evaluation of reservoir characteristics before and after dissolution. The dissolution effect on dissolution pore volume, dissolution ratio, pore diameter and connectivity can be quantitatively analyzed and calculated by the comparation of gypsum-salt and carbonate reservoirs before and after dissolution.

Experimental results: pore diameter. The pore diameter of combination of gypsum-salt rocks and carbonate rocks with different lithology before dissolution, ranges from 0.5 to 20 μm, with a big difference (Fig. 3.14). Among them, the pore

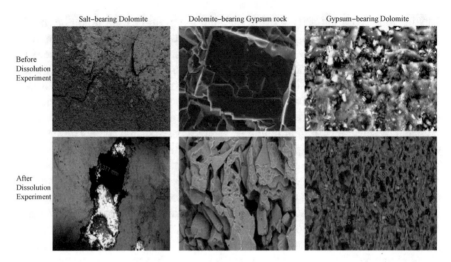

Fig. 3.14 Rock pores in symbiotic association of gypsum-salt and carbonate rocks before and after supergene karst to st

diameters of gypsum and dolomite before dissolution are relatively small, mainly intercrystal pores, with diameter of 1–2 μm, and surface porosity of 0.69 and 1.67% respectively. The pore diameter of gypsum-bearing dolomite is the second, mainly intercrystal pores and dissolution pores, with the pore diameter of gypsum of 1.5–5 μm, the pore diameter of dolomite of 1–6 μm, and the surface porosity of 2.05%. The largest pore diameter is developed in the salt bearing dolomite, with a surface porosity of 2.6%. The diameter of dissolution pores developed in gypsum is 2–10 μm, with the maximum of 17 μm. Microcracks are developed in the dolomite, with a pore diameter of 15–20 μm. The pore diameter of the reservoirs combined with gypsum-salt rocks and carbonate rocks increased significantly after the supergene karst test. The pore diameter of the salt bearing dolomite increased from 2 μm to 700–3000 μm, the pore diameter of the dolomitic gypsum increased from 1 μm to 6.6–21 μm, and the pore diameter of the gypsum-bearing dolomite increased from 0.582 to 23 μm. Pore connectivity. The pore connectivity of reservoirs combined with gypsum-salt and carbonate rocks in O_m of Ordos Basin is obviously increased by the supergene karstification. The samples of dolomitic gypsum before dissolution are developed with relatively low micro-porosity, relatively isolated pores, and poor connectivity. After 3D reconstruction from 7200 images obtained by 72 h of CT 3D scanning, a large number of connected pores with network distribution can be seen in samples of gypsum-salt and carbonate rocks after supergene karst. The pore connectivity in the samples decreases gradually from the surface to the interior by the research of pore volume and pore connectivity in samples with different pore diameters, based on the analysis of numerical reconstruction (Fig. 3.15). The pore connectivity of 10–50 um has the most obvious enhancement effect, and then the pore connectivity of 5–10 um is controlled by karst. Volume change. The dissolution amount and dissolution rate can be calculated respectively through the different weight of gypsum salt-carbonate

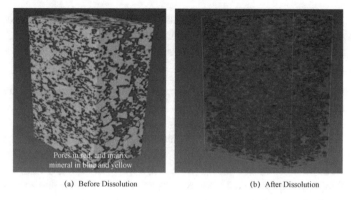

Pores in red, and matrix mineral in blue and yellow

(a) Before Dissolution   (b) After Dissolution

Fig. 3.15 3D distribution of pores in dolomitic gypsum before and after dissolution

combination samples before and after dissolution. Among them, the dissolution rate of salty bearing dolomite is the highest of 19.8%; the dissolution rate of dolomite-bearing gypsum is the second of 10.3%; the dissolution rate of dolomitic gypsum is between 8.2% and 6.9%; and the dissolution rate of gypsum-bearing dolomite is 6.3%. The gypsum rock is more easily dissolved than the dolomite in the combination of gypsum-salt and carbonate rocks, with the positive correlation between the dissolution amount and the content of gypsum-salt.

The concentrations of Ca^{2+}, SO_4^{2-}, and Cl^- in gypsum-salt rocks and carbonate rocks after dissolution are quite different as Ca^{2+} mainly comes from dolomite or anhydrite, SO_4^{2-} mainly comes from anhydrite and Cl^- comes from halite. After the experiment of salt-bearing dolomite, it can be seen that the main dissolution mineral is halite, with the highest concentration of Cl^- in 2777 mg / L and the concentration of Na^+ in 1815 mg / L. The concentration of SO_4^{2-} is the highest in the solution obtained from the experiments of dolomite-bearing gypsum, dolomitic gypsum and gypsum-bearing dolomite, and it decreases with the increase of anhydrite content. The concentration of SO_4^{2-} in dolomite-bearing gypsum and dolomitic gypsum is 497–1034 mg/L, which indicates that halite and anhydrite dissolve more easily than dolomite in gypsum-salt rocks and carbonate rocks under the real geological condition, with obvious pore enlargement effect and increase of porosity.

Therefore, the dissolution amount in symbiotic association of gypsum-salt rocks and carbonate rocks is positively correlated with the content of gypsum-salt under the closed system of atmospheric pressure, saturated CO_2 water (3–15%) and air temperature of 30 °C. The dissolution rates of salty bearing dolomite, dolomite-bearing gypsum, dolomitic gypsum, gypsum dolomite and gypsum-bearing dolomite are decreased in turn, with the porosity increases at least 6% after dissolution. The dissolution rate of salty bearing dolomite is the highest (19.8%), and that of gypsum-bearing dolomite is the lowest (6.3%). Therefore, the existence of large-scale and high-quality reservoirs in the O_m gypsum-bearing dolomite flat can be explained, as the porosity can be increased by 6–20% after karstification due to the development of favorable reservoirs composed of symbiotic association of gypsum-salt and carbonate rocks in O_m, such as salty bearing dolomite and dolomite-bearing gypsum. As a result, the karstification can be developed both in salty bearing dolomite and dolomitic gypsum-salt rock in evaporative tidal flat environment, and large-scale reservoirs of weathering crust karst can be developed.

② Gypsum-salt and carbonate rocks (combination type 1): dissolution simulation experiment with progressive burial

Aimed at the combination gypsum-salt and carbonate rocks in Cambrian, Tarim Basin, the experiment can be divided into closed system and open system. The closed system can be divided into five parts: No 1 crude oil, No. 2 deionized water, No. 3 salt rock solution, No. 4 gypsum solution, No. 5 CO_2 solution. The open system can be divided into three parts: No. 6 gypsum solution, No. 7 salt rock solution, No. 8 CO_2 solution. The samples are used to simulate the dissolution and reservoir formation of tight carbonate rocks under the condition of progressive burial with 1000–5000 m. The results show that (Fig. 3.16), the dissolution rate of rocks under acid condition (CO_2 solution) is the highest, and dissolution pores can also be formed in salt rock solution and gypsum rock solution. The pore diameter and porosity are obviously increased due to the obvious promotion of dolomite dissolution by CO_2 in solution under the environment of high temperature and high pressure. After dissolution,

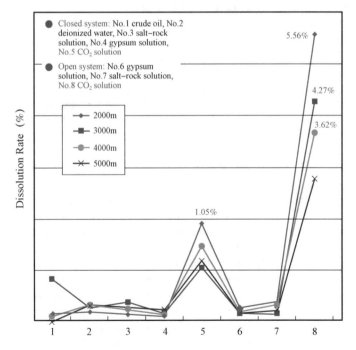

Fig. 3.16 Experimental results of pore changes before and after dissolution in progressive burial environment

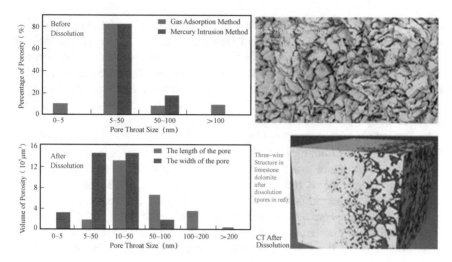

Fig. 3.17 Experimental results of pore change before and after dissolution in progressive burial environment

the physical properties of tight limestone dolomite become better, with the porosity from 0.75 to 4.2%, and 4.6 times in increase (Fig. 3.17). Therefore, the buried karst reservoir is worthy of attention as the physical properties of the reservoir can be improved by the buried karstification under appropriate conditions.

(4) The characteristics of the gypsum-salt rocks and carbonate reservoirs in Sinian-Cambrian of Sichuan Basin

The high-quality reservoirs composed of gypsum-salt rocks and carbonate rocks can be formed by multi-stage diagenetic transformation under the dissolution of salty fluid. It is favorable for the development of reservoir in the paleouplift and the superposition of paleouplift margin and grain beach in platform margin. Three sets of reservoirs composed of gypsum rocks and carbonate rocks are developed in Dengying Formation Member 2 and 4, and Longwangmiao Formation, Sinian-Cambrian, Sichuan Basin. The field outcrop reveals that regional uplifts generally develope in Late Sinian to Early Cambrian and Late Early Cambrian to Middle Cambrian. The karst reservoirs are mainly developed in Dengying Formaion, and three types of reservoirs with dolomite of grain beach, karst and hydrothermal are developed in Longwangmiao Formation.

① Reservoirs of Dengying Formation Member 2 and 4

Influenced by Chengjiang Movement and Tongwan Movement, Dengying Formation contacts angular unconformity with the underlying Doushantuo Formation and the overlying Qiongzhusi Formation of Cambrian respectively. Dengying Formation is divided into four sections from top to bottom. Under the influence of the first episode of Tongwan tectonic movement, the regional angular unconformity is developed between Dengying Member 2 and Member 3. At the end of the deposition of Dengying Member 2, the stratum of Dengying Member 2 is raised and suffered from weathering and denudation by the first episode of Tongwan tectonic movement. The karst reservoirs are formed by the development of a large number of solution pores and marked by cover of laterite weathering crust and calcium crust. Leaching and denudation with different degrees are formed in Dengying Member 4 under the influence of uplift in the second episode of Tongwan movement.

In two sets of reservoirs, Dengying Member 2 is dominated by platform margin facies and grain beach subfacies, and Dengying Member 4 is dominated by platform margin facies and grain beach subfacies. A platform sedimentary environment is developed in Dengying Formation of East Sichuan Basin, and transition to southeast and northeast as the facies of deepwater. The karst reservoirs are developed in the rugged paleogeomorphology of Dengying Member 2 (Fig. 3.18), caused by the regional uplift after early deposition of Dengying Formation. Such as the red weathering crust at the top of Dengying Formation Member 3 in the section of Laoqihe, the regional uplift movement between Dengying Formation Member 2 and 3 in the section of Luorendong, Ziyang, and the vadose pisolites fomed during the short-period weathering exposure of Dengying Formation Member 2 in the section of Yuduba. The karst reservoirs are developed in the rugged paleogeomorphology of Dengying Formation Mamber 4, caused by the regional uplift in Late Dengying Formation.

The grain beach are developed in Dengying Formation Basin, East Sichuan Basin, such as the high-energy grain beach in microbial mounds of Dengying Formation Member 4 of Yuduba, Wanyuan, and the thrombolit, oncolite and stromatolite dolimite in Dengying Formation Member 4 (Fig. 3.19a). The stromatolite dolimite

Fig. 3.18 Photos of weathering crust and dolomite of Dengying Formation in the section of Laoqihe and Yuduba **a** The internal boundary of the Dengying Formation Member 2 of Laoqihe section; **b** the weathering crust in the top of Dengying Formation Member 4 of Laoqihe section; **c** the dolomite bearing asphalt in the Dengying Formation Member 2 of Laoqihe section; **d** the dolomite bearing asphalt in the Dengying Formation Member 2 of Laoqihe section; **e, f** the oolitic sparry dolomite in the Yunduba section.

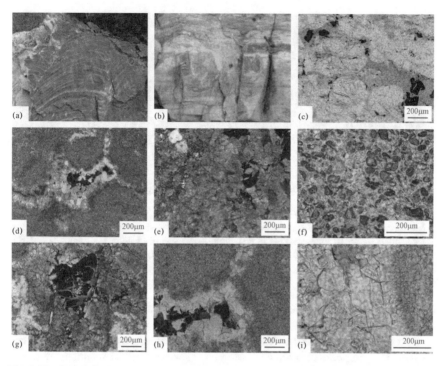

Fig. 3.19 Field and microscopic photos of Dengying Formation in East Sichuan Basin **a** Stromatolite of Dengying Formation Member 4 in Yuduba Section, field photo; **b** stromatolite dolomite of Dengying Formation in Liaojiacao Section; **c** sparite and grain dolomite, mainly composed of micrite dolomite sand, 4 × 10 times, casting thin section in blue, single polarized light, Guangongliang Section; **d** fine-grain dolomite, developed with intercrystal dissolution pores, 10 × 10 times, casting thin section in blue, cross-polarized light, Guangongliang Section; **e** residual particle silty dolomite, 4 × 10 times, casting thin section in blue, single polarized light, Guangongliang Section; **f** sparite and grain dolomite, mainly developed with micrite dolomite sand, 4 × 10 times, casting thin section in blue, single polarized light, Guangongliang Section; **g** crystal dolomite, half filled by asphalt in pores, 10 × 10 times, casting thin section in red, single polarized light, Laoqihe Section; **h** sparite and grain dolomite, 2 × 10 times, casting thin section in red, single polarized light, Longdonghe Section in Bashan; **i** sparite and grain dolomite, developed with intergranular pores and intragranular pores, 10 × 10 times, casting thin section in red, single polarized light, Longdonghe Section in Bashan

and botryoidal-lace dolimite are developed in Dengjing Formation of Liaojiacao, Pengshui (Fig. 3.19b), with the thickness of 500 m. The reservoirs of weathering karst are developed in the top of Dengying Formation. The oncolite dolomite, dissolution pores and asphalt are developed in the upper part of Dengjing Formation Member 4, Guangongliang Section, with the thickness of about 20 m. Dissolution poles partly filled with asphalt can be seen under the microscope (Fig. 3.19c, d, e, f). Dissolution dolomites with dissolution pores and cracks also can be seen in Dengying Formation by the microscope, with the porosity of about 15%, and residual asphalt can be seen

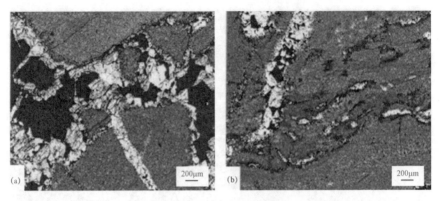

Fig. 3.20 Thin section of Well Guangtan 2, 6039.25–6039.32 m, Dengying Formation Member 4 under microscope. **a** Fracture micritic dolomite, fractures and caves are filled with dolosparite and black asphalt, 10×10 times, casting thin section, single polarized light; **b** stromatolite micritic dolomite, fractures are filled with dolosparite and black asphalt, 10×10 times, casting thin section, single polarized light

in the cracks (Fig. 3.19b). Intergranular pores can be found in Dengying Formation of Longdonghe Section, Bashan County, with porosity of about 4% (Fig. 3.19).

It also can be confirmed by drilling that the weathering crust reservoirs are developed in Dengying Formation, such as Member 2 and 4 in Well Li 1, with the reservoir space of secondary pores and dissolution pores, as well as large pores and relatively developed fractures. The logging of Dengying Formation is mainly lost circulation, with interpretation of water layer as 104.3 m/11 layers. The vertical depth of actual drilling in Dengying Formation is 833.5 m. Large pores and caves, as well as relatively developed fractures can be seen in the reservoirs. The reservoirs are developed with good quality, and mainly contained with water. The development of weathering crust karst reservoir is also confirmed by Well Guangtan 2. The results of core and thin sections show the development of black asphalt filling, which indicates that crude oil filling has occurred (Fig. 3.20).

② Reservoirs in Shilongdong Formation

A set of strata in limited carbonate platform facies and evaporite flat-shelf lagoon facies are developed in Shilongdong Formation of Sichuan Basin, with the intra-platform oolitic beach, sandy beach and karst reservoirs. The Shilongdong formation of Cambrian system in northeast Sichuan Basin is mainly composed of grain beach, and gypsum lagoons are locally developed, with reservoirs relatively developed. For example, grain dolomites with porosity of about 3% can be seen by the microscope in Bandenggou section; the reservoirs superimposed by grain beach and karst are developed in Shilongdong Formation, Laoqihecun section (Fig. 3.21). The reservoirs of grain beach are developed with the thickness of about 25 m, while oolitic dolomites with middle-thick layers in 17th layers are developed with the saccharoidal strata caused by strong weathering, and mound-beach is developed with the thickness of

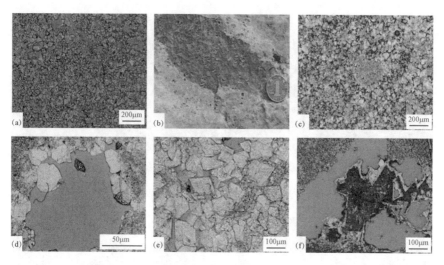

Fig. 3.21 Superimposed reservoirs of grain bank and karst in Shilongdong Formation of Laoqi-hecun Section. **a** Residual granular silty dolomite, 10 × 10 times, casting thin section in blue, single polarized light; **b** dolomite developed the karst cave, section photo; **c** residual granular silty dolomite, filled with asphalt in the intercrystalline pore, 10 × 10 times, casting thin section in blue, single polarized light; **d** residual granular silty dolomite, developed the karst cave, 4 × 10 times, casting thin section in blue, single polarized light; **e** residual granular silty dolomite, developed with intercrystalline karst pore, 20 × 10 times, casting thin section in blue, single polarized light; **f** residual granular silty dolomite, developed with cave, 4 × 10 times, casting thin section in blue, single polarized light

8.4 m; the oolitic dolomites with middle-thick layers in 24th layer are developed with the residual asphalt in intercrystalline pores, and the thickness of 2.7 m; thick oolitic dolomite is developped in the 25th layer, with worm-mold dissolution pores with a thickness of 2.7 m; the oolitic dolomites with thick layers in 28 and 29 layers are developed with worm-mold dissolution pores; the oolitic dolomites with thick layers in 29 layers are developed with intergranular pores of dolomite and intergranular solution pores of oolitic; the oolitic dolomites with thick layers in 30 layers are developed with worm-mold dissolution pores. The section is developed with relatively high porosity, ranging from 2.37 to 15.85%, with an average of 6.7%, and the permeability ranges from 0.0162 to 15.2 mD, with an average of 2.8 mD.

Regional uplifts are developed between the Early and Middle Cambrian, which are benefit to the development of karst reservoirs in Shilongdong Formation, such as the obvious weathering crust developed between Shilongdong Formation and Gaotai Formation in Taiyuan Section of Pengshui (Fig. 3.22).

Cave breccia is developed in Shilongdong Formation of Well Li 1 by core, with three times of stratigraphic leak during the drilling process, and 4 layers of water with the thickness of 17 m are interpreted, which may be caused by the weathering and denudation in the top strata (Fig. 3.23). The karst reservoirs in weathering crust of Shilongdong Formation are confirmed by the exploration of Well Li 1, with the

Fig. 3.22 Field photos of obvious weathering crust between Shilongdong Formation and Gaotai Formation in Taiyuan section of Pengshui

characteristics of high porosity and high permeability, a cumulative thickness of 30 m and a single layer thickness of 5–8 m in general. The average porosity and permeability of the reservoirs are 15% and 100 mD respectively.

③ Salt-related dolomite reservoirs in Cambrian

The gypsum-bearing strata of Longwangmiao Formation, Gaotai Formation and Xixiangchi Formation of Cambrian are mainly developed in the East-South Sichuan Basin, while the favorable reservoirs are developed in the area superposed by ancient uplift, the margin of ancient uplift and platform-margin grain beach. The granular dolomite reservoirs related to gypsum-salt are developed in Longwangmiao Formation and Xixiangchi Formation of Cambrian (Figs. 3.23 and 3.24). On the plane, the physical properties of the reservoirs near the ancient uplift are good, and those near the salt basin are relatively poor. Vertically, the reservoir physical properties become better from bottom to top, and the top of sequence boundary is the best. The favorable reservoirs are developed in the ancient uplift, the area superposed of margin of ancient uplift and platform-margin grain beach, which are controlled by sedimentary framework of ancient structure, grain beach and exposed surface with difference sequences as the distribution of Cambrian reservoirs is quite complex.

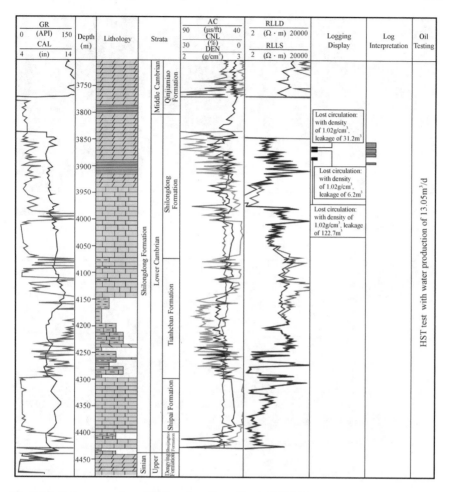

Fig. 3.23 Columnar section of Shilongdong Formation in Well Li 1

3.1.3 Superimposed Reservoirs of Microbialite

The research of microbialite mainly includes microbial limestone and microbial dolomite, which is a hot area for scholars at home and abroad. It refers to the the sedimentary rocks formed in situ by the capture and bond of clastic sediments through benthic microbial community, or the inorganic or organic inducing mineralization related to microbial activities. Microbial carbonate rocks are one of the most widely distributed microbialites, with the development age traced to ancient Archean, and mostly developed in Meso-Neoproterozoic, Cambrian and Ordovician (Fig. 3.25), including stromatolites, thrombolite and dendrites. At present, the good properties in microbial carbonate reservoirs of Proterozoic-Cambrian have been proved

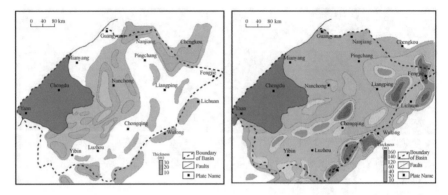

Fig. 3.24 Thickness of salt-related dolomite reservoirs in Longwangmiao Formation and Xixiangchi Formation, Sichuan Basin

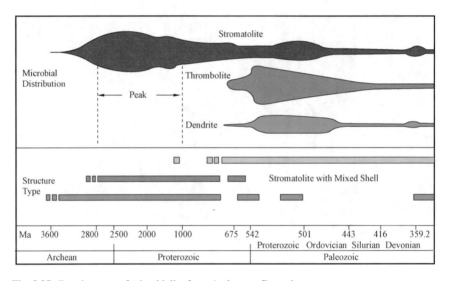

Fig. 3.25 Development of microbialite from Archean to Devonian

in Alabama of USA, eastern Siberia of Russia, South Oman salt basin, Santos Basin of Brazil, Kazakhstan, and Sichuan, North China, Tarim Basin in China. The reservoir accumulation mechanism of microbialite will be discussed in this paper by the examples of microbial carbonate rocks in Dengying Formation of Upper Sinian in the east Gaoshiti, Sichuan Basin and Xiaoerbulake Formation of Cambrian in Tarim Basin.

(1)   Microbialite of Dengying Formation in Upper Sinian, East Gaoshiti, Sichuan
       Basin

   ① Microbialite types

The biological development during the Dengying Formation in Sinian is obviously
different with that in Phanerozoic, with the aragonite sea environment of icehouse
caused by the relatively warm and dry paleoclimate in the whole Yangtze Region
(Xiufen et al. 2017). The microorganism build-ups with large number and different
forms are formed and preserved during the Dengying Formation. The main organisms deposited in Dengying Formation compose of bacteria and lower algae and lack
large-scale framework organisms. According to the special paleogeographic environment and sedimentary products of Dengying Formation, the deposition of Dengying
Formation can be defined by generalized reef, which mainly includes microbial reef,
microbial mound and lime mud mound. The microorganism (framework) reef refers
to the microorganism build-up with abundant microbial remains and wave-resistant
structures, with the development of framework pores. The microbial mound refers
to the microorganism buildup with pelletoid clot, lamination, microbial remains and
clot, foam layer, annular striation and snowflake structure. It develops with the
typical structure of wave-resistant structure, as well as the framework pores similar
to the microbial reef. The lime mud mounds refer to the microorganism buildups
composed of various micro-crystalline (crystalline) dolomites, with small structures
of microbial micro-crystalline, which can not be observed by naked eyes or magnifier.
   The carbonate rocks in Dengying Formation of Upper Sinian in Gaoshiti-Moxi
area are mainly composed of microbial dolomite. According to the observation of
core and microscope, the microbial dolomites of Dengying Formation are mainly
composed of stromatolite and microbial dolomite of non stromatolite. According to
the laminae form, the stromatolite dolomite can be divided into mound stromatolite dolomite and layered stromatolite dolomite, with a larger thickness and scale
in layered stromatolite dolomite. According to the structure, the non-stromatolite
dolomite can be divided into three types, including thrombolite dolomite, foam-layer
dolomite and micrite-pelletoid dolomite. The difference of main rocks in different
strata reflects the difference of the environment vertically.

Stromatolite microbial dolomite.

Stromatolite is first discovered in Archaeozoic (3.45 Ga) as the evidence of microorganism vital activity in Cryptozoic Eon, which is one of the main microbialite types in
Precambrian. It is widely distributed during the geological history due to the growth
under different water depth. As one of the main reservoir rock types of Dengying
Formation in central Sichuan Basin, stromatolite is mainly developed in the upper
part of intertidal zone and supralittoral zone, with various occurrences of lamination,
and developed in continuity, interruption and clutter. The dark lamination is mainly
composed of micrite or spherulite, and the bright lamination is mainly composed
of dolosparite. Fenestral structure is locally developed in research area (Fig. 3.26a–
d), and filled with dolosparite, quartz and asphalt in different degrees. The nearly

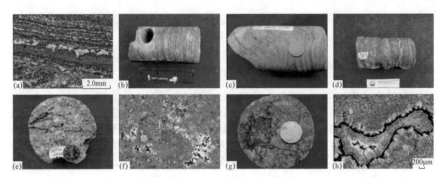

Fig. 3.26 Characteristics of microbialite in Dengying Formation of Gaoshiti area, Sichuan Basin. **a** Ordinary photo of laminated siliceous micrite dolomite, Member 4 of Dengying Formation, Well Gaoshi 21, 5322.3 m, single polarized light; **b** layered stromatolite, with interlaminar fenestral pores and intercrystalline pores filled by asphalt and quartz, upper Member 4 of Dengying Formation, Well Gaoshi 20, 5207.5 m, photo of core; **c** moundy stromatolite, with hard and tight rock due to the strong silicification, upper Member 4 of Dengying Formation, Well Gaoshi 21, 5321 m, photo of core; **d** moundy stromatolite, with interlayer karst caves filled by dolomite and asphalt, Member 2 of Well Gaoshi 2, 5397.7 m, photo of core; **e** thrombolite, Member 4 of Dengying Formation, Well Gaoshi 102, 5070.3 m, photo of core; **f** thrombolite, with pores filled by dolomite, asphalt and quartz, Member 4 of Dengying Formation, Well Gaoshi 102, 5169.68 m, cast thin sections in blue, single polarized light; **g** silicified non-stromatolite microbial dolomite, Member 4 of Dengying Formation, Well Gaoshi 21, 5315.37 m, photo of core; **h** thrombolite, developed with botryoidal-lace structure, Member 4 of Dengying Formation, Well Gaoshi 6, 5383.95 m, cast thin section in blue, single polarized light

round or elliptical spherulites are formed by the aggregation of very-fine aragonite or calcite particles through biochemical action, with the main diameter of 0.1–0.2 mm and fuzzy margin, as well as the dolomitization occurred in the early stage. The stromatolite is developed with the characteristic of rhythm, and a sedimentary cycle of microbialite is formed by the interbedding of layered stromatolite in deep gray and sand cutting/oolite in ligh gray, or foam-layered dolomite.

The mound stromatolite is developed as black hill-like protruding in the core, with the texture of hemispherical or box. The structures of the lamellar spherule and foam layer are associated with the mound stromatolite, Gaoshiti area, which indicate the appearance of biological colony of cocci in Sinian, Gaoshiti area. The mound stromatolites are mainly developed in the lower part of the carbonate sedimentary cycle in the carbonate platform, representing a relatively deep sedimentary product. As a result, it is generally considered that the mound stromatolites are formed in the intertidal zone. Under the condition of relatively deep water, the mound microbial build-ups are developed for the seek of more sunlight with the restriction of microbial population caused by the weak light. The interaction between the mound stromatolite and the layered stromatolite are quickly changed with the rhythmic changes of water environment.

Non-stromatolite microbial dolomite

The non-stromatolite microbial dolomite is a type of microbe rock with no obvious stromatolic structure. The macro characteristics of non-stromatolite microbial dolomite are difficult to study in the core observation. Therefore, in sieve residue logging data, it is often mistaken as doloarenite or granular dolomite. The microbe characteristics such as spherulites and foam-layer are found in the non-stromatolite microbial dolomite through microscopic identification. And the thrombolite, spherulite and foam-layer dolomite are developed in the non-stromatolite microbial dolomite of Dengying Formation, East Gaoshiti area.

Thrombolite, first proposed by Aitken, refers to the cryptalgal structure related to stromatolite without laminae structure. It is characterized by macro clotted structure, with irregular round at centimeter level, long strip branch or micro clotted structure in millimeter level. Modern thrombolite are usually developed in subtidal environment, deeper than that of stromatolites. The thrombolite are greatly developed in the eastern part of Gaoshiti area, and its thickness is only thiner to stromatolite (Fig. 3.26a–c).

Spherulite rock refers to the non-stromatolite microbial dolomite with microbial spherulite structure but no laminated structure. The spherulites is dispersively distributed, which is different from the microbial spherulites rock with dark laminae. The pure spherulites with small thickness are rare in the underground core of Gaoshiti area due to the large development of stromatolite and lamellar.

The foam-layer dolomite is a kind of non-stromatolite microbial dolomite with a microbe foam-layer like structure. The foam layers are developed as various shapes, such as irregular shape, honeycomb and lace. The foam-layer microbial dolomites may be formed in the intertidal zone with relatively strong hydrodynamic conditions as the terrigenous clastic and bacteria clastic mixed in the foam layers can be seen by microscope. The foam layers of underground rock cores in Gaoshiti area are isolated or connected to each other to form a network. The individual size is mainly 0.2–1 mm, with the edge of the individual in dark black due to the microbe micrite envelope. In addition, the formation of microbial mats (microbial membrane), the dark linear sediment formed by microorganisms, can be seen in the cavities between the foam layers. The cavities are developed as oval or long strip, and filled by dolomites present as equal-thickness cyclic-edge cements developed in multiple stages and generations, forming the so-called "botryoidal-lace structure" (Fig. 3.26). This kind of lace structures are well developed in Dengying Formation of the whole basin as the main reservoir structure.

② Characteristics of microbial reservoirs

Various types of microbial dolomite reservoir space are developed in Dengying Formation of Upper Sinian, East Gaoshiti, including cavity dissolved pores, intergranular dissolved pores, fenestral pores, mould pores of bacteria clastic and intercrystalline dissolved pores. Cavity dissolved pores are mainly developed in the interbedding of dolomite and stromatolite, and the formation mechanism is the cavity formed by the involution of microbial membrane (microbial mats). The "botryoidal-lace structure" is common formed by the equal-thickness cyclic-edge cements, and

Fig. 3.27 Reservoir space of microbial dolomite in Dengying Formation of Gaoshiti area, Sichuan Basin. **a** non-stromatolite foam-layer dolomite, developed with botryoidal-lace structure and cave, Well Gaoshi 6, Member 4 of Dengying Formation, 5373 m, ordinary thin section, single polarized light; **b** spherulite, mainly formed by fine-grained dolomite, with the development of intercrystal and intergranular dissolution pores, Well Gaoshi 7, Member 4 of Dengying Formation, 6288.9 m, cast thin section in blue, single polarized light; **c** stromatolite, fomred by micrite dolomites with dark laminae, bedding-parallel pores are developed, filled with granular quartz and asphalt, Well Gaoshi 101, Member 4 of Dengying Formation, 5497.2 m, cast thin section in blue, single polarized light

mixed with microbial membrane (Fig. 3.27a). Residual primary pores can be formed by the incomplete filling in small amout of cavities, while most cavities filled with cements are important reservoir space with the dissolution in buried period, or epigenetic karstification after structural uplift which reform the pore space. Intergranular dissolved pores are mainly developed in spherulites and thrombolite, and distributed around the spherulites and crumbs, with diameter generally less than 0.3 mm. The particles such as spherulites and crumbs are formed by microbial activities, and the content of organic matter is relatively high. The inside of the particles are mainly composed of micrite and organic matters, and it's surrounded by dolosparite. The residual dolomite can be seen in intergranular dissolved pores due to the selective dissolution of acid diagenetic fluid to the intergranular dolomite with the increase of temperature and pressure during the buried diagenetic period (Fig. 3.27b). Most of the fenestral pores belong to the primary pores formed in the early stage of carbonate deposition and are filled during the process of diagenetic evolution by the dolosparite calcsparite, quartz and asphalt (Fig. 3.27c). The favorable reservoir spaces can be formed by the dissolution of cements in the cavities due to the buried dissolution or supergene karstification during the late stage of diagenesis. The mould pores are mainly developed in the foam-layer dolomite. The individual bacteria clastic in the foam layer is easily mistaken for sand debris or oolite in hand specimens, but it can be identified under the microscope. The micrite "sclerotia" and cladding of microbial origin are mainly developed in bacteria clastics. The dissolution can be easily occurred in the bacteria clastics with micrite cladding in the epigenetic karstification, with the formation of mould pores. However, the mould pores are generally incompletely dissolved and lack connectivity due to the barrier action of the microcrystalline biological membrane in the foam layers.

Two sets of reservoirs are developed in Dengying Formation, Gaoshiti-Moxi area, which are Member 4 and Member 2 respectively. Member 4 of Dengying Formation can be divided into Upper Member 4 and Lower Member 4. The high-quality reservoir development area in the platform margin are mainly developed with the thickness of over 50 m, and mainly composed of microbial dolomites such as stromatolite and thrombolite, with the porosity of 6–12% and permeability of 0.5–5 mD. Large number of cavity dissolution caves, dissolution pores and fractures are developed in microbial dolomites with reduced thickness of reservoir in the inner platform area of East Gaoshiti area by the observation of core. The results of drilled wells show that the high-quality reservoirs are mainly developed in the Upper Member 4 of Dengying Formation, with the porosity of mainly 1–6% and the permeability of 10^{-5}-2 mD by the logging interpretation.

③ Controlling factors of reservoir development

The study of microbial dolomite reservoir in Dengying Formation of East Gaoshiti shows that the key controlling factors of reservoir development are sedimentary microfacies and diagenesis.

The types of original rocks are determined by sedimentary microfacies. The sedimentary facies analysis show that, the microbial reefs, mud mounds and grain beach are formed by the local uplift of intertidal zone in the platform, and the local development of microbial reefs and microbial beach are formed by the inner topographic relief in the epicontinental platform (Fig. 3.28). The microbial reefs are mainly composed of stromatolites mixed with thrombolite and form-layer dolomites, and the rock structures are favorable for the preservation of primary pores and the development of secondary pores caused by diagenesis transformation. As a result, the microbial reefs are the most favorable facies belt for reservoir. The grain beach can also form high-quality reservoirs due to the late transformations such as mould cavities in bacteria clastics and intergranular dissolution pores.

The reservoir space and pore structures of microbialite are mainly controlled by microbial structures. Taking stromatolite dolomite and non-stromatolite dolomite

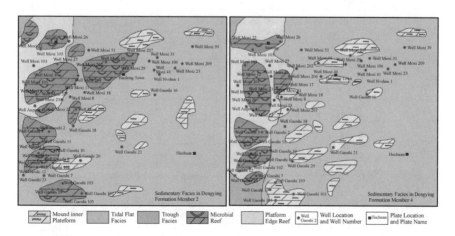

Fig. 3.28 Sedimentary facies of Member 2 and 4 of Dengying Formation in Gaoshiti-Moxi area, Sichuan Basin

as examples, the pore structures of them are quite different. The pore structures of stromatolite dolomite are mainly fenestral pores, with various shapes and filling degrees, and mainly distributed in bright color laminae. The pores of non-stromatolite dolomite are mainly dissolution pores, controlled by microbial particles and microbial mats, especially in foam-layer dolomite. The effective pores are generally developed in the cavities surrounded by microbial membranes, and the mould pores of bacteria clastics in the foam layers also contribute to the reservoir space.

The diagenesis on microbial reservoir performance can be divided into constructive and destructive effects. Based on a large number of petrological analysis, the diagenetic evolution sequence of dolomite in Dengying Formation of the study area can be summarized as follows: grow laminae of microbial mats (microbial membranes) / stromatolite structure-microbial dolomitization-microbial cladding and micritization-marine sparite cement filling-atmospheric fresh water dissolution in Tongwan period, formation of mould pores—sparite precipitation in mixed zone of seawater and freshwater-the first stage of silicification of siliceous fluid developed along laminae-the first stage of hydrocarbon filling along laminae—burial dissolution—granular cementation—pressure dissolution—the second stage of siliceous fluid, the horse-tooth quartz deposited in dissolved pores–the second stage of hydrocarbon filling–the third stage of siliceous fluid, the precipitation of granular quartz.The above diagenetic processes are occurred selectively in different rock types (such as stromatolite, spherulite, thrombolite and foam-layer dolomite) (Fig. 3.29).

The diagenesis analysis shows that microbial dolomitization, buried dissolution and supergene karstification play constructive roles in reservoir space. According to

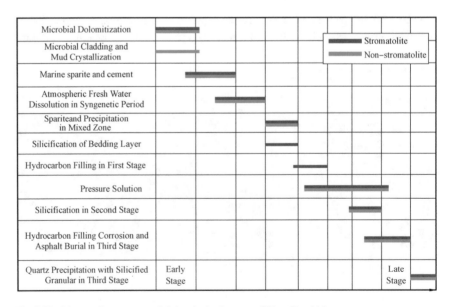

Fig. 3.29 Diagenetic sequence of dolomite in the core of East Gaoshiti

the identification of more than 50 thin sections in the study area, the surface porosity of the stromatolite microbialite can be increased in 2–3%, 3–6% and 5–6% respectively by the three diagenesis, and the surface porosity of the thrombolite, spherulite and foam-layer dolomite can be increased in 3–5%, 2–4%, and 4–6% respectively. Excessive dolomitization and silicification play a major role in destroying the reservoir space. According to the porosity statistics of the same type of microbial dolomite in Well Gaoshi 18 and Well Gaoshi 16, it is found that, excessive dolomitization can reduce the porosity of stromatolite microbialite by 2–3% and reduce the porosity of thrombolite, spherulite and foam-layer dolomite by 3–4%, while the silicification can reduce the porosity of stromatolite microbialite by over 10% and reduce the porosity of thrombolite, spherulite and foam-layer dolomite by 6–8%.

(2)   Microbialite in Xiaoerbulake Formation of Cambrian in Tarim Basin

① Microbialite types

Various types of microbial dolomites are developed in Xiaoerbulake Formation, with abundant changes of microbial structures. Among them, thrombolite dolomite, foam-layer dolomite and stromatolite dolomite are the typical rocks.

Thrombolite dolomite

The thrombolite dolomites are mainly distributed in the lower Xiaoerbulake Formation and the Member 1 of upper Xiaoerbulake Formation, with gray middle-thick layer. It is mainly composed of dark thrombolite and light cements on the macro level (Fig. 3.30a). Under the microscope, it can be seen as the filamentous algal laminar, which is composed of dark micrite dolomite, undulating and extending laterally, and partially intertwines with each other to form a closed "cladding" (Fig. 3.30b) or branches. The rare flat intergranular (dissolution) pores are formed by this unique structure, and distributed on both sides of the algal laminar (Fig. 3.30c).

Foam-layer dolomite

Foam-layer dolomites are mainly developed in Member 2 of Upper Xiaoerbulake Formation, with gray and white thick layers, usually developed with mound structure. The bedding-parallel dissolution pores and caves are widely developed on the surface and inside of the foam-layer dolomites by the field observation (Fig. 3.30d). Under the microscope, the foam-layer dolomite is made of elliptical foam cavities with different size, and developed like sponge. The cavity wall is composed of dark micrite dolomite. Unlike other types of microbialite, the algal framework of the foam-layer dolomite is very tight and stable with ubiquitous micritization, and its cavities are tightly bonded. As a result, pores do not develop between cavities, but within them, which are approximately elliptical in shape (Fig. 3.30e).

Stromatolite dolomite

Stromatolite dolomites are mainly developed in the Member 3 of Upper Xiaoerbulake Formation, with a thin-medium layer in gray to dark gray, as well as locally developed

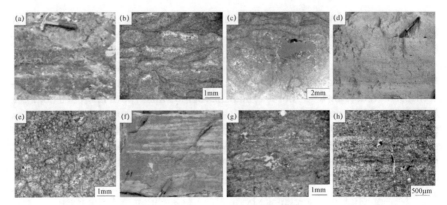

Fig. 3.30 Microbial structure of microbial dolomite in Xiaoerbulake Formation, Keping area, northwest margin of Tarim Basin. **a** thrombolite dolomite, with certain layered characteristics, poor continuity, Member 1 of Upper Xiaoerbulake Formation; **b** thrombolite dolomite, with the development of filamentous algal laminar, lateral extended, locally forming "cladding", Member 1 of Upper Xiaoerbulake Formation; **c** thrombolite dolomite, with flat intergranular dissolution pores developed on both sides of the laminar algal, Member 1 of Upper Xiaoerbulake Formation; **d** foam-layer dolomite, with the development of bedding-parallel dissolution pores, and pore diameter of mainly 3–5 mm, Member 2 of Upper Xiaoerbulake Formation; **e** foam-layer dolomite. The cavities in foam layers are elliptical with different sizes, and the pores are developed in the cavities, Member 2 of Upper Xiaoerbulake Formation; **f** stromatolite dolomite, with nearly horizontal laminae, Donggou 3 section of Xiaoerbulake, Member 3 of Upper Xiaoerbulake Formation; **g** stromatolite dolomite, with the structure of bedding-like bright-dark laminae. The pores are developed between the dark laminae. Member 3 of Upper Xiaoerbulake Formation. **h** mud crystal power dolomite, with flat algal clastic arranged along the layers, and fenestral pores developed, Lower Xiaoerbulake Formation

in the Member 2 of Upper Xiaoerbulake Formation. The morphology of stromatolite dolomites is transited from wave in Member 2 of Upper Xiaoerbulake Formation to weak wave and layer in Member 3 of Upper Xiaoerbulake Formation (Fig. 3.30f). The laminae structure in bright and dark can be seen under the microscope. The dark laminae are composed of bond structure with strong micritization and laminae formed by cyanobacteria spherules, with weak continuity. Some of laminae can be scattered into patches, which are easy to be mistaken as clot structures; while the bright laminae are composed of crystal power dolomites with coarser grain size and cleaner surface. The pores of stromatolite dolomites are mainly developed with framework pores (Fig. 3.30g), and developed between the dark laminae with the flat shape. The long axis of framework pores is parallel to the direction of the laminae, which may be related to the laminae contraction in early diagenetic period.

② Characteristics of microbialite reservoirs

According to the study of lithofacies paleogeography and sedimentary facies, the microbial carbonates of the Xiaoerbulake Formation are widely developed in the platform margin and the platform interior (Fig. 3.31). According to outcrop sections, drilled wells and seismic data, the mound-beach facies in the platform are mainly

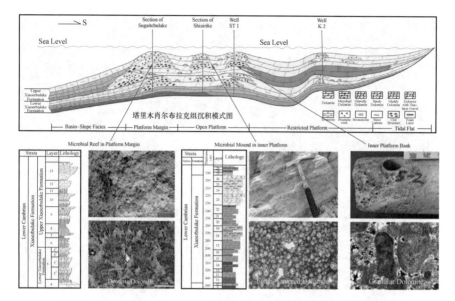

Fig. 3.31 Sedimentary models and microbial mound types of Xiaoerbulake Formation

developed in north of Bachu-Tazhong, northwest of Keping Uplift and northeast of Tabei Uplift, the microbial build-ups of platform margin are mainly developed in Lunnan-Gucheng, with the distribution area of reef and mound-beach facies of 10.5 × 10⁴ km², which has the basis of large-scale reservoir development.

The thrombolite dolomites are mainly developed in the mound base which has relatively low energy, with poor reservoir properties and average porosity of 2.01%, through the detailed anatomy of the sedimentary sequence in the microbial mounds of the Xiaoerbulake Formation (Fig. 3.32). The form-layer dolomites and grain dolomites are mainly developed in the mound cores and mound caps with higher energy. The reservoir properties is relatively high, and the porosity is 5.47–13.51%, which constitutes the main part of the reservoir.

③ Controlling factors of reservoir development

The pore types of microbial carbonate in Xiaoerbulake Formation mainly include fabric selective pores such as interstitial pores of clots, foam-layer framework pores, intergranular (intra) pore, and partly non fabric selective pores such as solution pores, solution cracks and solution caves and pores. The formation of reservoir is mainly controlled by dominant facies belt and short-term exposure corrosion.

Microbial mounds and beaches are the material basis of reservoir development, and the key to reservoir formation is the penecontemporaneous atmospheric fresh water dissolution caused by the short-term drop of sea level (Jing et al. 2014). A large number of uneven dissolution ditches and pits are formed due to the exposed erosion and leaching developed in the section of Upper Xiaoerbulake Formation. At the same

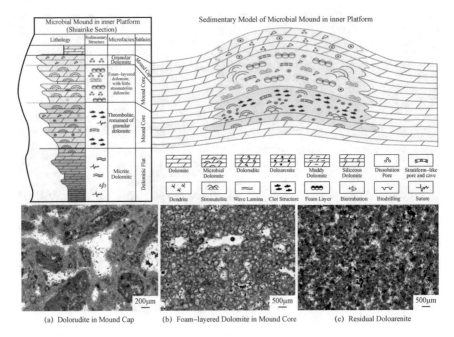

Fig. 3.32 Microbial mound structure and reservoir development of the Xiaoerbulake Formation

time, the crescent cements and suspended cements related to seepage zone of atmospheric fresh water (Fig. 3.33a, b), and the cementation "unconformity" phenomenon caused by the local dissolution of undersea fibrous cements and filling of granular cements in the later burial period can be seen under the microscope (Fig. 3.33c). The reservoir rocks transformed by short-period exposed dissolution have the characteristic of porphyritic cathodoluminescence of medium brightness (Fig. 3.33e, f). Meanwhile, the carbon and oxygen isotope values of the reservoir near the dissolution interface are slightly negative shift and fluctuated significantly compared with the seawater in the same period (Fig. 3.34), which is significantly different from the surface karst reservoir and buried hydrothermal dissolution reservoir formed by long-period exposed dissolution.

A large number of pores are still left in the Cambrian microbial carbonate reservoirs after a long period of burial diagenesis, and the retentive diagenesis plays an important role. Early dolomitization is penecontemporaneous microbial dolomitization and refluxing dolomitization in shallow burial period. After dolomitization, the anti-compaction and anti-pressolution are improved in microbialite (Fig. 3.35), which is beneficial to the maintenance of pores formed in quasi-contemporaneous period (Haoyuan et al. 2018). At the same time, the gas released by microorganisms, especially methane ancient bacteria, not only helps to overcome the kinetic barrier bacteria formed by dolomite at low temperature, but also maintains the pre-existing

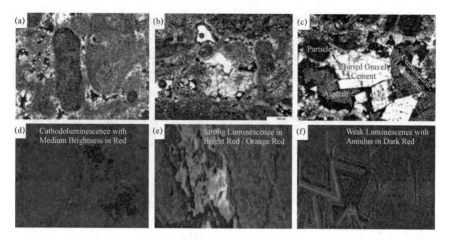

Fig. 3.33 Micrographs of Xiaoerbulake Formation reservoir in Cambrian of Tarim Basin. **a, b** Characteristics of seepage zone with suspended cements, Well Shutan 1; **c** the cement unconformity, Xiaoerbulake Formation, Well Shutan 1; **d** Medium-bright cathodoluminescence in red, short-period exposed dissolution, 35th-layer in Xiaoerbulake Section; **e** strong luminescence in bright red / orange-red, weathering crust karst, epigenetic atmospheric water, Well Yaha 5; **f** girdle with weak luminescence in dark red, buried hydrothermal dissolution, Well He 4

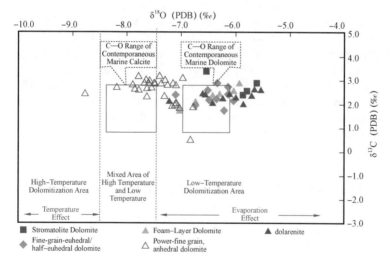

Fig. 3.34 Characteristics of carbon and oxygen isotopes in Xiaoerbulake Formation

pores in reservoir with the inhibition of cementation by reducing permeability and preventing the fluid convection in the two-phase fluid system formed in sediments due to the generation of a certain amount of methane gas.

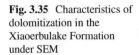

Fig. 3.35 Characteristics of dolomitization in the Xiaoerbulake Formation under SEM

3.2   Accumulation Mechanism of Deep Clastic Rocks

The effective reservoirs can be formed as favorable accumulation for oil and gas in deep and ultra-deep clastic rocks such as Kuqa Depression, Junggar Basin and Bohai Bay Basin, due to the better physical properties compared with other types of rocks, which is caused by the long-term burial, compaction and dissolution, under the influence of the tectonic and sedimentary environment (Zhong et al. 2008). The deep clastic reservoirs in China are characterized by wide distribution, long time span, great difference in physical properties and various pore types. Nowadays, it is generally believed that dissolution, burial mode, abnormal high pressure, gypsum-salt effect, and clay membrane to the preservation of primary pores all have great influences on the formation of deep and ultra-deep effective reservoirs. The reservoir formation mechanism of deep and ultra-deep clastic reservoirs in China can be summarized into three points, through the collection and arrangement of a large number of previous research.

3.2.1   Three Mechanisms of Pore Retention

(1)   Favorable pore preservation under the condition of low geothermal gradient and rapid deep burial in the late stage after long-period shallow burial

The low geothermal gradient has a certain preservation effect on the primary pores, and also makes the deep-buried reservoirs in the period of a large number of secondary pores. The higher the geothermal gradient is, the stronger the diagenetic strength is, and the geothermal gradient has a significant control effect on the porosity of sandstone. The effective reservoirs are deep buried with the slow attenuation rate of sandstone porosity in the area with low geothermal gradient. Tarim Basin is a typical "cold basin" with the average geothermal gradient of about 20 °C /km, and an effective reservoir distribution range of 4500–6000 m. It is beneficial to the preservation of primary pores in the quartz sandstone with the point contact of particles due to the weak pressolution in quartz grains even when the buried depth is over 5700 m, which is caused by the strong anti-compaction of quartz sandstone due to the low geothermal gradient. At the same time, the ground temperature is about 100 °C where is deeper than 5000 m due to the low geothermal gradient, so it's in the maturity period of organic matters, which is beneficial to the dissolution of cements, matrixes and mineral particles formed in the early stage and improvement of sandstone porosity. Therefore, the porosity of sandstone of Tarim Basin can still be as high as 20% when the burial depth is 5000–6000 m.

The simulation experiment shows that the reservoir physical properties can be better preserved by the rapid burial. The sandstone reservoir of the Cretaceous Bashijiqike Formation in Kuqa foreland basin is an important gas-producing zone of the West to East Gas Transmission in China, and it can be a representative for the favourite reservoirs under strong compaction and buried depth over 5000 m. The reservoir was deposited 130 Ma ago, and sedimentary and diagenetic evolution process of the reservoir can be divided into two stages: (1) The early stage of long-period shallow burial in 130–23 Ma ago. Under the slow and stable compaction in early diagenetic stage, the stratum subsided to 2000–3000 m. (2) The later stage of rapid burial from 23 Ma to now. Under the strong compaction in middle-late diagenesis, stratum rapidly subsided to about 9000 m. The fractures are obviously developed in sandstones due to the tectonic stress. The reservoir space types of deep and ultra-deep sandstones in the Bashijiqike Formation of lower Cretaceous, Dabei area, Kuqa Depression are mainly composed of intergranular pores (residual primary intergranular pores and intergranular dissolved pores), feldspar (or debris) intragranular dissolved pores and micropores, with the development of microfractures, and the main pore combination is residual primary intergranular pores-dissolution pores-micropores, due to the special burial processes. It can be confirmed that the porosity of the quick burial sandstone in the later stage is 5% different from that of the normal burial sandstone by the numerical simulation in ultra-deep layers of Kuqa, Tarim Basin (Fig. 3.36). The primary pores are preserved for a long time in southern margin of Junggar Basin due to the slow-shallow burial in the early stage of sedimentation and weak intergranular compaction, and they are preserved more effectively in the late stage because of intergranular fluid remained in the reservoir caused by the short rapid burial, and the inconspicuous destruction on the sandstone by the chemical action. Therefore, the relatively high-quality reservoirs are still developed, with porosity up to 15.6% even where the buried depth is more than 6000 m (Fig. 3.37).

(2)   Good preservation of ultra-deep reservoir space filled by hydrocarbon in early stage

The influence of hydrocarbon filling in early stage on diagenesis, especially on compaction in deep and ultra-deep clastic reservoirs is also worthy of attention. The primary pores can be preserved by the reduction of compaction caused by the inhibition to the cementation of quartz and carbonate minerals by draining the fluids

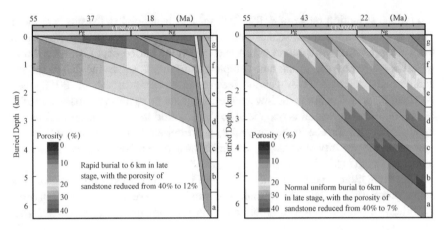

Fig. 3.36 Simulation experiment of porosity change in clastic rocks of Kuqa Depression, Tarim Basin

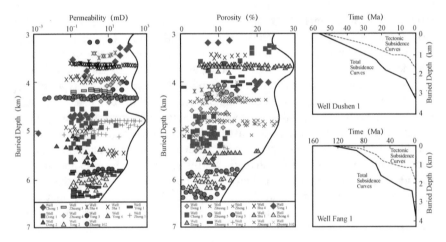

Fig. 3.37 Relationship between physical properties of rapidly buried clastic reservoirs and buried depth in Junggar Basin

from the pores with the early-stage injection of hydrocarbon. The primary intergranular pores are preserved in fine-grained debris-quartz sandstone reservoirs of Well Tazhong 4, Tarim Basin, with a depth of 3650.3 m, due to the inhibition of diagenesis by hydrocarbon injection, with a porosity of 21%. The porosity of fine-grained sandstone reservoirs of 12% in Well Tazhong 17 is small, with the phennmenon of quartz secondary outgrowth, strong diagenesis with no hydrocarbo filling (Plate III a, b). The difference in porosity between the oil-bearing sandstone and the oil-free sandstone in Carboniferous of Tazhong is 9% (Fig. 3.38). This phenomenon is widely distributed in deep and ultra-deep clastic reservoirs in China, such as Ordos Basin,

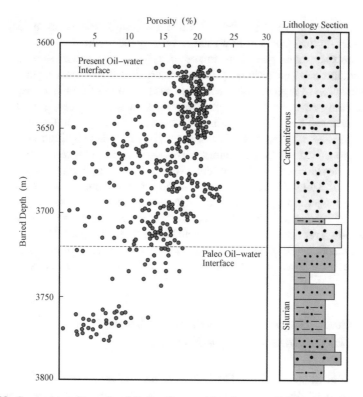

Fig. 3.38 Comparison of porosity of Carboniferous with early-stage oil filling and early-stage non filling in Tazhong area, Tarim Basin

Tarim Basin, Junggar Basin, Huanghua Depression and so on. The ultra-deep reservoir space can be well preserved by early-stage hydrocarbon filling, mainly for three reasons: first, the early-stage hydrocarbon filling changes the original geochemical environment and water wettability of the rocks; second, it inhibits the diagenesis happened in the original water–rock medium; third, it accelerates the dissolution of feldspar and carbonate minerals.

(3) The primary pores can be well preserved by the protective effect of clay membrane

Clay membrane can effectively increase the anti-compaction of rocks and inhibit the formation of authigenic minerals as to preserve pores. Clay membrane is widely developed in deep and ultra-deep clastic reservoirs in China, such as Silurian and Devonian of Tarim Basin, Jurassic and Cretaceous of Songliao Basin, Paleogene and Neogene of Dongpu Depression, Triassic and Jurassic of Sichuan Basin, Triassic of Ordos Basin. The mineral composition are mainly smectite, illite, kaolinite, chlorite, quartz, calcite (Plate III c, d), and can be divided into two types: primary mineral and autogenetic mineral. Under medium and low compaction, pores in deep buried sandstone reservoirs can be well preserved by clay membrane.

Chlorite clay membrane is an important type of clay membrane, and widely developed in deep and ultra-deep clastic reservoirs. It has great significance for the preservation of high porosity, with the characteristics of thin layer, equal thickness and continuous growth, and grows around the edge the rocks, mostly in radial vertical particles. The Xujiahe Formation reservoirs developed in the west Sichuan Depression is a typical tight sandstone reservoir, with relatively high-quality reservoirs developed in the tight environment. Through a variety of research methods, it is found that the chlorite around-edge cement developed in Upper Member 4 and Lower Member 2 of Xujiahe Formation, plays a good role in the preservation of primary pores. The contact strength between particles of chlorite-lined rocks is relatively low, generally with point-to-line contact. The authigenic chlorite has a close relationship with the preservation and evolution of sandstone reservoir space. Sun Quanli et al. Think that authigenic chlorite is mainly contributed in preventing other cements precipitate in the pores, inhibiting compaction and promoting particle dissolution.

3.2.2   Quick Dissolution of Sandstone Under High Temperature and High Pressure

The formation of deep high-quality reservoirs in China are almost related to dissolution, mainly due to the dissolution of intergranular carbonate cement by organic acid and inorganic acid (carbonic acid formed by carbon dioxide) produced by organic matters, followed by the secondary dissolution of feldspar and debris. Dissolution is the most common mechanism in the development of high-quality clasitc reservoirs in deep and ultra-deep strata, with different degree of dissolution in different regions. In the past experiments, it was considered as the underdevelopment of ultra-deep pores with the restriction of atmospheric pressure, and the dissolution temperature was less than 120 °C. The high-pressure and high-temperature simulation results of 180 °C and 53 Mpa show that the dissolution rate with the temperature over 150 °C (Fig. 3.39) increases by 2–3 times (Fig. 3.40). For example, the ultra-deep reservoirs in Cretaceous of Tarim Basin are still well developed (Fig. 3.41), and the water formation in Well Keshen 7 under acid frauturing, which proves the development of high-quality clastic reservoirsin with buried depth of more than 7900 m.

3.2.3   Pores Improved by Fractures and Faults

The fractures and faults can be easily formed by the rock failure under the structural stress caused by the increase of rock brittleness in ultra-deep strata. The fluid percolation space can be increased by the great improvement of reservoir property in

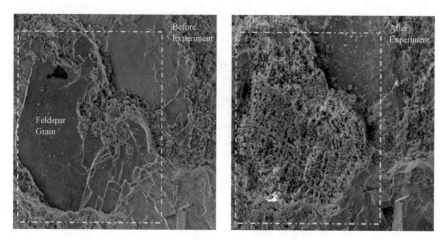

Fig. 3.39 Comparison of dissolution in clastic rocks under different temperature and pressure

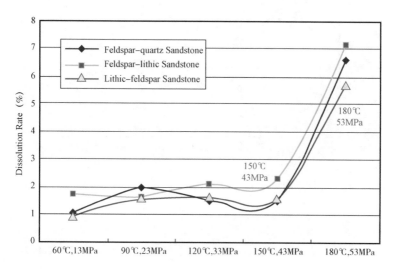

Fig. 3.40 Simulation experiments on dissolution mechanism of clastic rocks under different temperature and pressure

tight reservoirs caused by fractures and faults. For example, fractures are developed in Well Dabei 202 (Fig. 3.42, Plate III h, i), with no transformation in sandstone reservoirs, the permeability of 0.132 mD and gas production of 110×10^4 m^3 / d. On the contrary, the fractures are not developed in Well Yinan 2, with the permeability of 0.013 mD and low production.

The large-scale reservoirs are developed in Cretaceous of Kuqa Depression, with delta sandbodies stacked vertically and connected horizontally. The reservoirs are developed with the thickness of 200–400 m and an area of 1.8×10^4 km^2. The

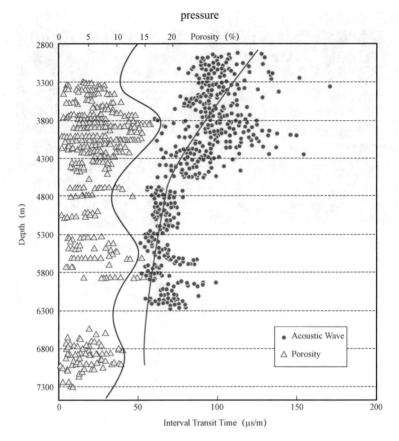

Fig. 3.41 Corresponding relationship between porosity and acoustic wave of Cretaceous in Kuqa area

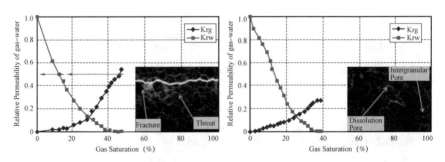

Fig. 3.42 Comparison of development of reservoir fractures between Well Dabei 202 (left, 0.132 mD) and Well Yinan 2 (right, 0.013 mD)

effective reservoirs can be developed in the buried depth of 6000–8000 m, with the porosity of 5–10%, due to the preservation of primary pores caused by the long-term shallow burial and rapid deep burial in late stage, as well as the development of structural fractures.

3.3 Accumulation Mechanism of Deep Volcanic Reserviors

Volcanic reservoir is an important type of reservoir. The pores of volcanic reservoirs can be easily preserved as the porosity is less affected by the compaction depth, as the strong anti-cpmpaction caused by the harder framework of volcanic rocks compared with other rocks and the diagenesis of volcanic rocks mainly in the form of condensation and consolidation. The volcanic reservoirs can become the major reservoirs as the reservoir capacity of volcanic rocks is better than that of sedimentary rocks where the burial depth is greater than a certain depth. Nowadays, volcanic reservoirs have been found in more than 300 basins and blocks of more than 20 countries, such as Kurosaka gas reservoir in Niigata Basin in Japan, Scott Reef oil and gas reservoir in Browes Basin in Australia, Lapa oil and gas reservoir in Neuquen Basin in Argentina, and Jatibarang oil and gas reservoir in Jawa Basin in Indonesia (Caineng et al. 2008; Zihui et al. 2008). The volcanic reservoirs have been found in various oil and gas basins in China since 1950s, such as Wucaiwan Depression in Junggar basin, Tahe Area in Tarim Basin, Santanghu Basin, Xujiaweizi Fault Depression in Songliao Basin et al.

3.3.1 Distribution of Continental Volcanic Reservirs and Two Types of Effective Reservoirs

Volcanic rocks have been developed in most of the petroliferous basins in China, with the large distribution and thick strata, which are mainly distributed in four basin groups and three sets of strata. As shown in Fig. 3.43, the four basin groups are developed with a total area of 36×10^4 km^2, including the eastern basin group (7×10^4 km^2), the northern Xinjiang basin group (9×10^4 km^2), the Tarim Basin (13×10^4 km^2) and the Sichuan-Tibet basin group (7×10^4 km^2). Generally speaking, the volcanic rocks are distributed in the strata of Carboniferous-Permian, Jurassic-Cretaceous and Paleogene. The volcanic reservoirs are mainly composed of basic-intermediate volcanic rocks, such as basalt, andesite, dacite, as well as intermediate-acid volcanic rocks, such as rhyolite and tuff.

Compared with sedimentary rocks, the effective reservoirs of volcanic rocks are mainly as two types of weathering crust type and primary type, which are more complex in reservoir type, more variable in physical properties and more heterogeneous. The formation conditions and distribution rules of different types of volcanic

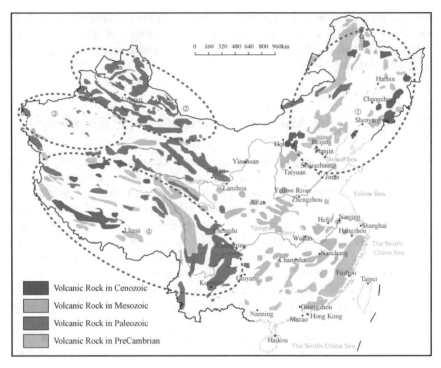

Fig. 3.43 Age and distribution of continental volcanic rocks in China ① eastern basin group; ② northern Xinjiang basin group; ③ Tarim basin group; ④ Sichuan-Tibet basin group

favorable reservoirs vary greatly in different basins or regions. The volcanic reservoirss are mainly developed as weathering crust reservoirs in the west, and the primary reservoirs in the east. Statistics show that the porosity and reservoir quality can be greatly improved by certain secondary alteration, no matter what the lithology is Fig. 3.44.

The weathering crust of volcanic rocks refers to the different combination of minerals, structures and reservoir characteristics formed by sedimentary hiatus, weathering leaching and alteration. It can be identified by weathered clay layer, argillaceous fillings of oxidation environment in faults, iron oxide lining in self-broke fractures in micro photos and geopetal structure. The effective reservoirs can be formed by volcanic weathering crusts in various lithologies under the control of weathering and erosion, with the reservoirs space mainly of secondary dissolution pores and fractures. The effective reservoirs of volcanic weathering crust are mainly developed within the depth of 550 m below the unconformity surface with the underdevelopment of faults (Fig. 3.45), and mainly developed within the depth of 1100 m below the weathering crust with the development of faults. The research of the wells which drilled the complete Carboniferous volcanic weathering crust in Northern Xinjiang (Jinghong et al. 2011) shows that five layers of soil layer, hydrolytic zone, dissolution zone, disintegration zone and parent rock are developed

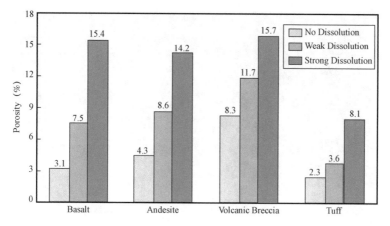

Fig. 3.44 Statistics of alteration porosity of volcanic rocks paulopost alteration

from top to bottom in volcanic weathering crust (Fig. 3.46). The soil layer is the product of strong alteration in volcanic rocks, with the form of soil and poor reservoir performance, mostly composed of secondary minerals; the hydrolytic zone is the product of strong alteration in volcanic rocks, which is mainly composed of fine particles and mudstone, and developed with poor reservoir performance; the dissolution zone is the product of relatively strong alteration of volcanic rocks, which is mainly composed of the volcanic fragments, with the development of secondary pores and fractures; the disintegration zone is the product of medium alteration of volcanic rocks, mainly composed of relatively large volcanic fragments, with the development of secondary pores and fractures, where fractures and pores are often filled or semi-filled;the parent rock is the unalterer volcanic rocks. Statistical physical properties show that the soil layer and hydrolytic zone are not effective reservoirs, accounting for 31% of the thickness of weathering crust. The dissolution zone and disintegration zone are effective reservoirs, accounting for 69% of the thickness of weathering crust. The effective reservoirs of volcanic weathering crust are mainly distributed in high position, slope and fault in low position, controlled by the palaeogeomerphology landform, faults and period of weathering and leaching.

The reservoir space of primary reservoir can be divided into primary pores and primary fractures. The primary pores are composed of the stomata, formed by volcanic materials ejected to the surface, and the residual pores which are not completely filled by the amygdale, intercrystal pores and inter-volcanic breccia pores. The stomata are formed with different sizes due to the escape of volatile components during the eruption of magma, which are largely contained in the magma during the overflow process. Pores in amygdale refer to the space left after the filling with secondary minerals, which are formed by the dissolution of filling minerals. Inter-crystal pores refer to the pore produced by mineral crystallization between crystals, which are related to the degree of mineral crystallization. The higher the degree of crystallization, the more the pores are developed. The primary fractures refer to the

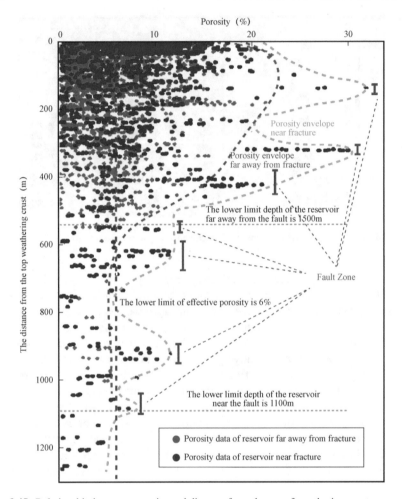

Fig. 3.45 Relationship between porosity and distance from the top of weathering crust

burst fractures and contraction fractures formed by volcanism and diagenesis. Fractures can be the main seepage channels and part of the reservoir space of volcanic rocks.

3.3.2   Reservoir Space Type Under the Control of Four-Stage Inner Structure

The four-stage inner structure in volcanic rocks is developed with volcanic body facies-reservoir permeator unit-pore structure, which directly control the type and reservoir space of volcanic reservoirs. The gas–water relationship and the macro

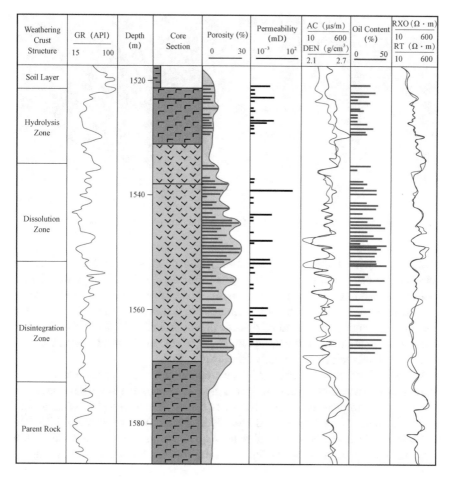

Fig. 3.46 Structure and characteristics of volcanic weathering crust

distribution of the reservoirs are controlled by volcanic body; the reservoir development are controlled by volcanic facies; the macro mobility of the reserves, well pattern and well distance are controlled by reservoir-permeator unit; the micro availability of the reserves and the seepage law of the reservoirs are controlled by the pore structure (Fig. 3.47).

Volcanic body, which is generally developed close to the deep-large faults, controls the gas–water relationship and the macro distribution of the reservoir with a large scale of several kilometers, in the form of central mound and fracture-layer. The degree and scale of the reservoir space are fundamentally determined by the differences of lithology and lithofacies conditions. The primary pores are developed during the formation of volcanic rocks, including the supporting space generated by differential consolidation of magma in the flow process, the space left by the gas–liquid inclusions lossing in the melt after the diagenesis process, as well as the contraction

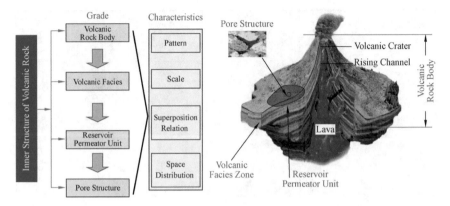

Fig. 3.47 Volcanic reservoirs controlled by four-stage inner structures of volcanic rocks

fractures and joints formed during the condensation process of magma. The reservoir performance can be directly affected by the lithology, as the lithology of volcanic rocks can determine the type of primary pores. The viscosity and brittleness of rocks are gradually increased from basic lava, intermediate lava to acid lava. The pores and fractures are largely developed in rhyolite and andesite, and mainly composed of stoma. Intergranular pores and intragranular pores are mianly developed in volcanic breccia, and microcracks are mainly developed in tuff.

Volcanic lithofacies is controlled by the eruption type, transportation mode, location environment state of volcanic materials, and their formation. It can be divided into six basic types: eruptive facies, effusive facies, extrusive facies, eruption-sedimentary facies, volcanic channel facies and buried hill facies. At present, the division of volcanic lithofacies at home and abroad is quite inconsistent. Taking the central eruption volcano as an example (Fig. 3.48), the volcanic lithofacies can be roughly divided into: the eruptive facies formed on the surface, including the overflow facies, the eruptive facies and the extrusive facies. volcanic neck facies formed from the surface to the magma chamber or volcanic source area, with rock products of lava, pyroclastic lava and pyroclastic rock; the subvolcanic facies formed about 3 km below

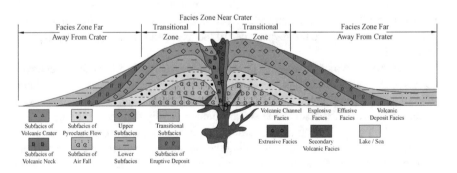

Fig. 3.48 Models of volcanic lithofacies

the surface, with rock products of lava, breccia lava and breccia. The volcanic sedimentary facies formed on the surface, with rock products of extrusive rock, sedimentary pyroclastic rock, pyroclastic sedimentary rock and sedimentary rock. Different pore types are developed in different lithofacies and subfacies (Table 3.3). Reservoir physical properties of different subfacies in the same lithofacies may vary greatly, as the great differences in rock construction and structure of facies and subfacies, which control the combination and distribution of primary and secondary pores and fractures. The isolated stomas and inter volcaniclastic pores are mainly developed in reservoir space of volcanic channel facies. the inter volcanic breccia pores, stomas and dissolution pore-cave-fracture are developed in the volcanic eruption facies. The primary stomas and contraction fractures of lava are developed in the volcanic eruption facies, and the secondary pores are mainly structural fractures. the fractures, dissolution pores and intercrystal pores are mainly developed as the reservoir space in subfacies of the central zone of the extrusive facies. Generally speaking, volcanic reservoirs space can be divided into four types: stoma type, intergranular pore type, dissolution pore type and fracture type.

The stomas are formed by the volatilization effusion of volcanic lavas. The high-quality reservoirs can be developed by the connection of structural fractures and contraction fractures due to the strong anti-compaction of volcanic rocks. The porosity reduction can be made by later diagenesis such as albite filling in hydrothermal period, kaolinization of matrix in epigenetic period, secondary quartz overgrowth and carbonate filling in burial period. Taking the Yingcheng Formation of Well Shengshenping 1 as an example, it is a development well, located in Shengping-Xingcheng structure of northern slope of Xujiaweizi fault depression, the south of Wangjiatun gas field, Daqing, with an actural drilling inclined depth of 3700 m and a daily gas production of $(22.1–55.5) \times 10^4$ m^3. The characteristics of "hourglass shape" is developed in the longitudinal direction of volcanic reservoir. A high-quality reservoir of stomata rhyolite is developed in subfacies of upper effusive facies, with a thickness of 516 m, followed by the lower rhyolite and poor tight rhyolite in the middle. On the plane the reservoirs are widely developed near the volcanic craters, and reduced away from the craters.

Intergranular pores are developed by the accumulation and compaction of volcaniclastic materials, with the relatively high-quality reservoirs by the late-stage fracture connection. On the contrary, the pores will be lost bacause of the late-stage diagenesis such as mechanical compaction, secondary quartz overgrowth and carbonate cementation in buried period. Generally, the characteristics of "anti rhythm" is developed in intergranular pores in the reservoir, with the good properties in the upper reservoir, and poor in the lower. The reservoirs are widely developed near the volcanic craters on the plane, and reduced away from the craters. Taking the discovery well of Xushen 1 in Xushen Gasfield as an example, the daily gas production is 22×10^4 m^3. Yingcheng Formation can be divided into four sets of strata from bottom to top: Yingcheng Formation Member 1, Yingcheng Formation Member 2, Yingcheng Formation Member 3 and Yingcheng Formation Member 4. Yingcheng Formation Member 1 is an acid volcanic section overlying the angular unconformity, with main lithology of crystalline tuff in dark gray and black-gray, rhyolite in

Table 3.3 Volcanic lithofacies of petroliferous basins in China (Caineng et al. 2008)

Facies	Subfacies	Formation depth	Rock type	Formation method	Output state
Volcanic channel facies	Volcanic crater-volcanic Neck	Magma area in surface, or volcanic source area	Lava, pyroclastic lava and pyroclastic rock	The volcanic edifice is denuded, exposing the filling of thevolcanic channel	Oval, circular and polygonal on the plane
	Sub-volcanic rock	3 km below surface	Lava, breccia lava and breccia	Contemporaneous or late invasion	Rock bed, rock wall, rock stock, rock branch and cryptoexplosive breccia body
	Cryptoexplosive breccia		Cryptoexplosive breccia	The magma rich with volatile components intrudes into the broken rock zone and produces underground eruption	Tube, layer, vein, branch, fork and fracture filling
Eruptive facies	Airfall	Surface	Massive Pyroclastic Rock	Volcanic eruption product	Multi-facies system composed of volcanic clastics, surrounding clastics and vapour
	Thermal Base -Surge	Surface	Tuff with wave structure, volcanic breccia tuff		
	Pyroclastic Flow	Surface	Alloying Pyroclastic Rock	Volcanic eruption	Clastic flow accumulation of volcanic eruption
Effusive facies	Upper, middle and lower part	Surface	Various Lavas	Products of volcanic eruption and flood	lava flow, rock quilt; rope, column, slag, etc
Extrusive facies	Center, transition, and margin	Surface	Lava and Breccia Lava	Products extruded to surface, such as lava from the volcanic neck	Rock needles, rock plugs, rock bells and domes
Volcanic eruptive-sedimentary facies	Sedimentary pyroclastic rocks, tephra deposits containing extraclast, reworked tephra deposits	Surface	Effusive rock, pyroclastic rock, pyroclastic sedimentary rock and sedimentary rock	Sedimentary products of gap period and low tide period in volcanic eruption	Continental and basin facies; bedded and lenticular deposits, etc

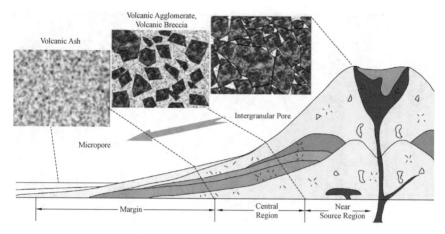

Fig. 3.49 Volcanic reservoir with intergranular pores of Well Xushen 1

gray and gray-white, and variegated volcanic breccia. The interbedding of sand and mudstone with tuffaceous matter are developed in Yingcheng Formation Member 2. The basic-intermediate volcanic rocks are developed in Yingcheng Formation Member 3, mainly composed of andesite in dark purple and dark gray, andesite tuff in purplish red and grey-green, and altered diorite porphyrite. Variegated conglomerate with huge thick and interbedding of sandstone and mudstone with tuffaceous matter are mainly developed in Yingcheng Formation Member 4. The volcanic lithofacies in turn are splashing subfacies, pyroclastic flow subfacies, themal base-surge subfacies and airfall subfacies. The reservoir is mainly vertically distributed in the volcanic clastic rocks with the splashing subfacies and pyroclastic flow subfacies of the upper eruptive facies, and becomes worse downward. Horizontally, the reservoir is mainly developed in the eruptive facies near craters, and the degree of development away from the craters decreases (Fig. 3.49).

The dissolution pores are formed by the dissolution of acid water such as organic acid and CO_2. The organic acid has a strong ability to dissolve feldspar, carbonate and matrix, while the ability of CO_2 acid water to dissolve feldspar and volcanic ash is weak. The geological reserves of natural gas in Carboniferous volcanic gas reservoir of Kelameili Gasfield is 1053.34×10^8 m^3, with the daily gas production of 30×10^4 m^3. The major gas reservoir is volcanic reservoir, mainly developed in basalt, tuffaceous breccia, feldspar porphyry and ignimbrite. The feldspar porphyry of subvolcanic rocks is mainly developed in area of Well Dixi 18. The eruption mode is dominated by the central type, supplemented by the fracture type, and dominated by the eruptive facies and subvolcanic facies. The reservoir near the crater on the plane is better than that far away from the crater. The reservoir properties gradually reduce from the top of the upper interface to the bottom as the loss of dissolved materials and the stability destruction of the parent rocks caused by weathering and erosion in the vertical direction (Fig. 3.50). The pores and fractures are developed at the top of the subvolcanic rock, with small density, and average porosity of 20% and

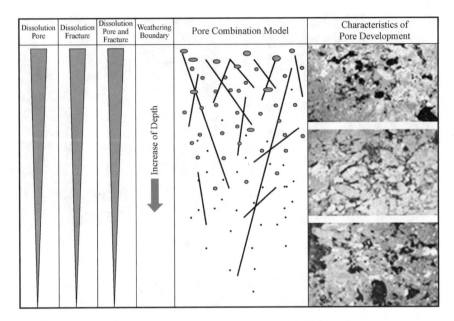

Dissolution Pore	Dissolution Fracture	Dissolution Pore and Fracture	Weathering Boundary	Pore Combination Model	Characteristics of Pore Development

Fig. 3.50 Volcanic reservoir with dissolution pores in Well Dixi 18

the daily gas production of 30×10^4 m^3. The development of the pores and fractures decreases and the density increases in the middle part of subvolcanic rock, while the secondary pores are not developed in the lower part. Therefore, the reservoirs are relatively developed near the upper interface of the rock body and the volcanic channel.

The development of secondary pores are controlled by constructive transformation in later stage. The primary pores in volcanic rocks are not always connected with each other, so the permeability is low. The secondary pores formed by transformations such as later tectonization, weathering and leaching, dissolution and devitrification are essential for the formation of volcanic reservoirs. On the one hand, the reservoir spaces are increased, on the other hand, the connectivity and permeability of the reservoirs are increased, so the reservoir performance are increased.

(1) Tectonic stress action. A large number of fractures and faults with different scales are developed by structural activities, and the reservior spaces are formed which can improve the reservoir performance. At the same time, the connectivity of various types of pores are strengthened by the structural fractures, and the reservoir performance is further improved. (2) Weathering and leaching. The porosity and permeability of the volcanic rocks exposed to the surface can be significantly improved by the weathering and leaching of the atmosphere and fresh water in the eruption gap or after the volcanic activities. (3) Dissolution. The secondary pores can be formed by the dissolution in diagenesis, and the connectivity between pores can be also strengthened. To the volcanic rocks without the development of primary pores, the key factor to the formation of effective reservoirs is the dissolution in late

stage. For example, the main pores of rhyolitic lava tuff and rhyolitic (crystalline) tuff are developed from dissolution pores, and the secondary porosity is as high as 74.1 and 81.7%. (4) Devitrification. The diagenesis in volcanic reservoir can be developed with the increase of burial depth and temperature, and sometimes it may lead to the loss of some primary pores. Such as the conversion of biotite to clay minerals, which will lead to the shrink of the pores due to volume expansion of minerals. For example, the felsic minerals with globular, radial or fibrous are formed by the devitrification of rhyolitic glasses, during the process of diagenetic evolution of the deep spherulite rhyolite reservoirs of Yingcheng Formation, Songliao Basin. A quite large number of micropores in rocks can be developed by the decrease of net volume after the formation of crystalline minerals, as the density of rhyolitic glasses is less than that of feldspar and quartz.

References

Baomin, Zhang, Liu Jingjiang. 2009. Classification and characteristics of karst reservoirs in China and related theories. *Petroleum Exploration and Development* 36(1): 12–39.

Caineng, Zou, Zhao Wenzhi, Jia Chengzao, et al. 2008. Formation and distribution of volcanic hydrocarbon reservoirs in sedimentary basins of China. *Petroleum Exploration and Development* 35 (3): 257–271.

Guangyou, Zhu, Zhang Shuichang, and Liang Yingbo. 2006. Formation mechanism and distribution prediction of high-quality marine reservoir in deeper Sichuan Basin. *Petroleum Exploration and Development* 33 (2): 161–166.

Haoyuan, Yu., Cai Chunfang, Zheng Jianfeng, et al. 2018. Effect of microbial structure on the porosity of microbial dolostone - a case study of Xiaoerbulake Formation in Keping area of the Tarim Basin. *Petroleum Geology and Experiment* 40 (2): 233–243.

He, Youbin, Jinxiong Luo. 2010. Lithofacies palaeogeography of the late Permian Changxing age in middle and upper Yangtze region. *Journal of Palaeogeography* 12(5): 497–514.

Hu Mingyi, Fu., Cai Quansheng Xiaoshu, et al. 2014. Characteristics and genetic model of karst reservoirs of Ordovician Yingshan-Yijianfang Formation in Halahatang area, North Tarim Basin. *Geology in China* 41 (5): 1476–1486.

Jinghong, Wang, Jin Jiuqiang, Zhu Rukai, et al. 2011. Characters and distribution patterns of effective reservoirs in the Carboniferous volcanic weathering crust in Northern Xinjiang. *Acta Petrolei Sinica* 32 (5): 757–766.

Jing, Zhang, Zhang Baomin, and Shan Xiuqin. 2014. Controlling effects of paleo-climate and paleo-ocean on formation of carbonate reservoirs. *Petroleum Exploration and Development* 41 (1): 121–128.

Ling, Li., Tan Xiucheng, Zeng Wei, et al. 2013. Development and reservoir significance of mud mounds in Sinian Dengying Formation, Sichuan Basin. *Petroleum Exploration and Development* 40 (6): 666–673.

Schroder, S., B.C. Schreiber, and J.E. Amthor. 2010. A depositional model for the terminal Neoproterozoic & Early Cambrian Ara Group evaporates in south Oman. *Sedimentology* 50 (5): 879–898.

Wei, Yang, Xie Wuren, Wei Guoqi, et al. 2012. Lithofacies paleogeography, favorable reservoir distribution and exploration zone in Cambrian-Ordovician of the Sichuan Basin. *Acta Petrolei Sinica* 33 (Supplement 2): 21–34.

Xiufen, Zhai, Wang Zecheng, Luo Ping, et al. 2017. Characteristics and origin of microbial dolomite reservoirs in Upper Sinian Dengying Formation, eastern Gaoshiti area, Sichuan Basin, SW China. *Natural Gas Geoscience* 28 (8): 1199–1210.

Yigang, Wang, Wen Yingchu, Hong Haitao, et al. 2004. Exploration target of gas reservoir in deep oolitic beach of Feixianguan Formation of Triassic in Northeast Sichuan. *Natural Gas Industry* 24(12): 5–9.

Zhong Dakang, Xiaomin Zhu, and Hongjun Wang. 2008. Analysis on the characteristics and formation mechanism of high quality clastic reservoirs in China. *Science China (Seri. D)* 38(Supplement 1): 11–18.

Zihui, Feng, Shao Hongmei, Tong Ying, et al. 2008. Controlling factors of volcanic gas reservoir property in Qingshen gas field, Songliao Basin. *Acta Geologica Sinica* 82 (6): 760–768.

Chapter 4
Formation and Distribution of Deep Large Oil and Gas Fields

How can large-scale oil and gas accumulation be developed in deep reservoirs, and how does oil and gas accumulate and distribute with various types of hydrocarbon kitchen and hydrocarbon phases, as well as the superimposition and transformation of multi-stage tectonic movements under the deep geological conditions. As the important ways to the formation of deep large oil and gas fields, the hydrocarbon accumulation of deep oil and gas crossing tectonic periods are developed with two different mechanisms, such as the accumulation mechanism of deep liquid hydrocarbon reservoir in the Tarim Basin and the accumulation mechanism of gas reservoir in the Sichuan Basin. The theory of the "exploration golden zones" in the deep superimposed basin is developed, with the main sources control ment in deep oil and gas accumulation, which the distribution of oil and gas field was controlled by mainhydrocarbon kitchens, and the oil and gas enrichment was controlled by paleouplifts, slopes, paleoplatforms and paleofractures. The theory is based on the formation of large-scale oil and gas fields controlled by tectonic-lithofacies paleogeography setting reservoir-forming factors such as hydrocarbon kitchen, reservoir, passage system and caprock, which revelaed a good explration prospect.

4.1 Formation Conditions of Deep Large Oil and Gas Fields

With complex mechanisms in oil and gas accumulation, the enrichment degree is uneven by various facters (Jinxing et al. 2007; Du Jinhu et al. 2013). Generally speaking, the oil and gas accumulation in the basin are controlled by the basic characteristics of hydrocarbon accumulation conditions, such as source rocks, reservoirs, cap rocks, source-reservoir-cap configuration, traps, migration and preservation, and the temporal and spatial matching relationship. The formation of deep-large oil and gas fields was controlled by the combination of four factors: the sufficiency of hydrocarbon kitchens, the scale of reservoirs, the effectiveness of migration system and the

© Petroleum Industry Press 2021
S. Hu and T. Wang, *Deep-Buried Large Hydrocarbon Fields Onshore China: Formation and Distribution*, https://doi.org/10.1007/978-981-16-2285-4_4

sealing of caprocks. This research analyzes the formation conditions of deep-large oil and gas fields by taking deep carbonate study as a case.

4.1.1 Sufficiency of Source Kitchens

The marine hydrocarbon-bearing strata in the superimposed basin have the characteristics of developing in old age, spanning a long time and bearing many hydrocarbon layers. The development of large-scale oil and gas accumulation is difficult with complex conditions because the deep source rocks are general in the high-overmature stage. At the same time, the oil and gas accumulations in deep strata are developed with the characteristics of crossing tectonic periods and long-term hydrocarbon losses and destruction, which are caused by the destruction and transformation of multi-stage complex structural movements. As a result, it is particularly important to develop sufficient hydrocarbon kitchens.

There are two meanings of sufficient hydrocarbon kitchens: the first is the effective allocation of source rocks and reservoirs, that is the source control theory. The relatively high degree of oil and gas filling in the long geological history process can only be developed by the large-scale hydrocarbon accumulations formed in the large-scale high-quality reservoirs, which were supplied with source nearby or had the high-efficiency migration system composed of faults and unconformities. Secondly, the guarantee for the formation of large oil and gas fields are the high-efficiency large-scale source rocks and effective timing of hydrocarbon generation and expulsion. Source rocks with large scale is very important (Zhaoyun et al. 2004) which was the material basis of large-scale oil and gas accumulation after a long-term large-scale hydrocarbon losses. The distribution of Anyue Gasfield in the Sichuan Basin is obviously controlled by the hydrocarbon generation center of Deyang-Anyue Rift, with the area of 3×10^4 km^2, and the thickness of 360–580 m, three times of the adjacent area. The organic carbon abundance in the central area is more than 2%, twice of the adjacent area. The gas generation intensity in the central area is $(60–100) \times 10^8$ m^3/km^2, two–three times of the adjacent area. The formation of oversized gas field in Anyue was guaranteed by such a high-quality hydrocarbon kitchen. The effective timing of hydrocarbon generation and expulsion was a sufficient guarantee for the formation of large oil and gas fields, especially the combination of late-stage generation-expulsion and traps. The effectiveness of oil and gas accumulation was determined by the massive hydrocarbon generation and structural ceasing in the late stage. The main reservoir formation periods of the discovered large oil and gas fields in the continental large-scale superimposed basin are mainly formed after Cretaceous, expecially in the Paleogene, whether they are the ancient marine carbonate strata or the Meso-Cenozoic continental clastic strata (Table 4.1). The preservation and discovery of large number of large-scale oil and gas fields in China under such a complex geological background is mainly attributed to forming oil and gas reservoirs in late stage, which can avoid the damage of multi-stage tectonic movement to the greatest extent.

Table 4.1 Statistical table of late-stage accumulation in major large oil and gas fields in China' onshore

Basin	Large oil and gas fields	Reservoirs	Type of oil and gas	Main accumulation period
Tarim Basin	Kela 2, Dina No. 1, Dina No. 2, Dabei 2, Kekeya	K, E	Dry Gas	E-N
	Tahe No. 1, Tahe No. 2, Lunnan, Sangtamu	O, T, J	Oil	N-Q
	Tazhong, Hetianhe	O-C	Oil	K-E
	Yingmai 7, Yingmai 2	K-E, O	Oil	E-N
Sichuan Basin	Luojiazhai, Dukouhe, Puguang, Longgang, Yuanba et al	P-T	Dry Gas	N-Q
	Tiandong, Datianchi, Wolonghe, Fuchengzhai	C	Dry Gas	N-Q
	Moxi, Gaoshiti, Longnvsi, Hebaochang	Z-$\in$	Dry Gas	N-Q
	Weiyuan, Ziyang	Z	Dry Gas	N-Q
	Guangan, Moxi, Zhongba, Xinchang, Pingluoba	T_3x	Dry Gas	N-Q
Ordos Basin	Sulige	P	Wet Gas	K
	Jingbian,Yulin,Zizhou, Wushenqi,Shenmu	C-P	Dry Gas	K
Junggar Basin	Hutubi	E	Dry Gas	E
Qiadam Basin	Sebei No. I, Sebei No. II	Q	Biogas	Q

4.1.2 Scale and Effectiveness of Reservoirs

The exploration in deep strata has been one of the succeeding fields of oil and gas exploration strategy with the increase of degree and difficulty in exploration and development of middle-shallow strata. However, more requirements for the scale of deep carbonate reservoir are put forward bacause the cost of deep oil and gas exploration and development is much higher than that of medium and shallow strata. The scale and effectiveness of deep carbonate reservoirs are the necessary conditions for the formation of large oil and gas fields. At present, the discovered large-scale oil and gas reservoirs in China are basically defined. In spite of the diversity of types, the carbonate reservoirs can be divided into two types: reservoirs controlled by facies and reservoirs controlled by diagenesis. The geological backgrounds of large scale different reservoirs are proposed through the analysis of carbonate reservoirs in the Tarim, Sichuan and Ordos basins. The research shows that the large-scale carbonate reservoirs controlled by facies are mainly developed in three sedimentary environments such as evaporation platform, carbonate slope and platform margin.

The large-scale carbonate reservoirs controlled by diagenesis are developed with complex factors and uncertainty under the control of the scale of reservoirs pre-existing and hydrothermal solution. The diagenetic reservoirs controlled by local tectonic movements are generally developed with large scale. The recognition of the geological background of large-scale reservoirs has important guiding signifi-cance for the evaluation of deep carbonate exploration field. Recently, three types of reservoirs are worthy of attention in deep carbonate reservoirs exploration of Tarim, Sichuan and Ordos basins: reef-beach reservoir, karst reservoir and dolomite reservoir (Table 4.2).

(1)   Development conditions for the large-scale reservoirs controlled by facies

Reservoirs controlled by facies can be divided into sedimentary dolomite reservoirs and reef beach reservoirs. The deep large-scale carbonate reservoirs are mainly devel-oped in evaporation platform, margin of rimmed platform and carbonate gentle slope through the exploration of Tarim, Sichuan and Ordos basins.

The sedimentary dolomite reservoir developed in the evaporation platform were large-scale developed. For example, the gypsum-dolomite reservoirs in Upper Maji-agou Formation (Member 5_{1-4}) of the Ordos Basin were developed as a typical Sabha dolomite reservoir, and fine-grained dolomite reservoirs in Middle Majiagou Formation (Memeber 5_{5-10}) were developed as a typical seepage-reflux dolomite reservoir (Fig. 4.1), which were exposed to the unconformity in the west of Jingbian Gasfield, and the main part of Jingbian Gasfield is located in the gypsum-dolomite transitional zone. The inter-salt and under-salt dolomite reservoirs in Middle-Lower Cambrian of Tarim Basin, and dolomite reservoirs in Jialingjiang Formation and Leikoupo Formation of Sichuan Basin are all related to the evaporation platform in arid climate.

The high-energy facies belts are the basis for the large-scale reservoirs, mainly including reef and beach, in which the scale of beach bodies is larger than reef bodies. Two types of reef-beach are developed in 12 strata of three major marine basins in China (Fig. 4.2). The area of reef-beach on the platform margin is $(13-16) \times 10^4$ km^2, and the area of beach inner platform is $(23-30) \times 10^4$ km^2, which can be as the material basis for the large-scale reservoirs. Superimposed reformation with karst is the key to the formation and maintenance of large-scale reservoirs. The high-quality dolomite reservoirs of Longwangmiao Formation developed with karst reformation, with the matrix porosity of 3%–5%. The high-yield wells with production of million cubic meters are mainly concentrated in the section of dissolved pores. The dissolved pores decrease from the higher to lower part of the paleouplift, and the physical properties of the reservoir gradually deteriorate. Therefore, the superimposed karstification reformation on the high-energy facies belts is an important field for large-scale high-quality dolomite reservoirs.

(2)   Geological conditions for the development of large-scale diagenetic reservoirs

Diagenetic reservoirs include dolomite reservoirs of burial and hydrothermal refor-mation, and buried hills (weathering crust) karst reservoirs. The controlling factors

Table 4.2 Development potential and main controlling factors of large-scale marine carbonate reservoirs in China

Reservoir type				Development potential and main controlling factors of large scale reservoir
Reservoir controlled by facies	Reef-beach Reservoir	Reef-beach Reservoir in margin of rimmed platform		It has the development potential of large-scale reservoir, and the large-scale reef and grain beach are the main controlling factors
		Reef-beach Reservoir in margin of rimmed platform and inner platform		The scale of reservoir is uncertain, and controlled by barrier types, barrier continuity, platform type, water depth and geovnerphic faetures
		Reef-beach Reservoir in slope of platform		It has the potential of large-scale reservoir, and the large-scale grain beach in carbonate slope platform is the main controlling factor
	Dolomite Reservoir	Sedimentary dolomite reservoir	Sabha dolomite	It has the potential of large-scale reservoirs, and the large-scale gypsum-dolomite, reef-mound and dolomite in reef-beach of carbonate evaporation platform are the main controlling factors
			Reflux seepage delomite	

(continued)

Table 4.2 (continued)

Reservoir type				Development potential and main controlling factors of large scale reservoir
Reservoir controlled by diagenesis		Dolomite reservoir with burial and hydrothermal reformation	Buried dolomite	It has the potential of large-scale reservoirs, and mainly controlled by the scale of preexisting reef-beach reservoirs. Buried dolomite from limestone and dolomite of non reef-beach facies may be an important supplement.
			Hrdrothermal dolomite	As the scale of hydrothermal dolomite reservoir was controlled by the types, scale and action time of hydrothermal fluid, as well as the scale and distribution range of faults, fractures and permeability layers, it has much uncertainty
	Karst Reservoir	Interlayer karst reservoir		Related to regional tectonic movements and generally developed in large scale
		Bedding karst reservoir		Associated with karst reservoir of buried hill, expanding the exploration scale of karst reservoir developed as an important supplement

(continued)

Table 4.2 (continued)

Reservoir type				Development potential and main controlling factors of large scale reservoir
		Buried hill (weathering crust) karst reservoir	Buried limestone hill	Related to local tectonic movements and generally developed in large scale
			Crust in dolomite	
		Karst reservoir controlled by fractures		Related to local tectonic movements and generally not developed in large scale

of large-scale reservoirs were more complex as the reservoirs were not controlled by facies, which increased the uncertainty of large-scale reservoirs development.

Effective configuration of fracture-cave karst reservoirs and source fractures is also a type of large-scale reservoir. For example, the large-scale fractures and caves in Ordovician interlayer karst reservoirs of the Tarim Basin were developed with karst superposition and transformation of fault systems (Fig. 4.3), with the distribution area of interlayer karst reservoir in slope over 2×10^4 km^2. The fractures and caves can be formed by the dissolution of groundwater in early-stage faults, and connected by late-stage faults; and the large-scale karst caves are distributed along the fault zone. The exploration practice shows that the success rate of drilled wells in Halahatang fault zone is 96%, while that in non-fault zone is less than 50%. The result shows that the fault zone in the interlayer karst limestone reservoirs is a favorable area for the development of large-scale fractures and caves, as well as the deployment of high-efficiency wells.

The dolomites formed by diagenetic transformation can be divided into buried dolomite and hydrothermal dolomite, which are the focus of the research on deep large-scale reservoirs, and may be one of the main controlling factors of the large-scale development in deep high-quality reservoirs. The hydrothermal dolomite reservoirs in the lower Yingshan Formation of Tazhong Uplift are widely distributed in phyre, lenticular and quasi-layered forms along faults, fractures and permeable layers. The intercrystal pores, intercrystal dissolved pores and dissolved pores of hydrothermal dolomites in Lower Yingshan Formation of Zhonggu 9, Gulong 1 and Gucheng 6 are developed, with porosity of 10%–12%. The daily gas production of Gulong 1 is 10067 m^3, and the daily gas production of Gucheng 6 is 264234 m^3. The reservoirs are developed with poor lateral continuity and strong heterogeneity, which are hard to develop in large scale. The hydrothermal activities are active in Tazhong 3, Tazhong 12, Tazhong 18, Tazhong 45, Tazhong 80, Tazhong 162 and the outcrop of Yingshan Formation, Liuhuanggou section, with the formation of hydrothermal

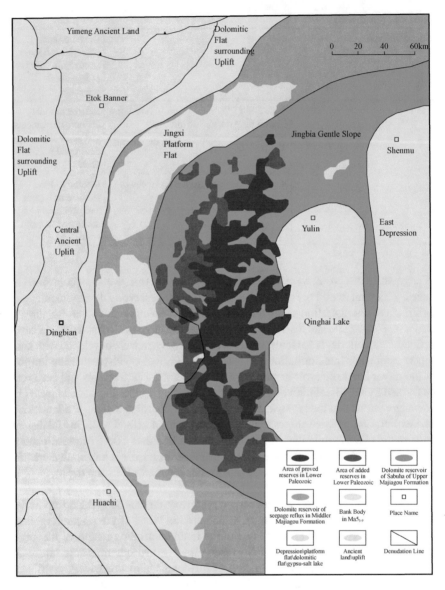

Fig. 4.1 Distribution of reservoirs with favorable facies belts of Majiagou Formation in the Ordos Basin

dolomite, the development of large caves, and the pores of millimeter or centimeter level, which are partly filled with hydrothermal minerals. The dolomite reservoirs of grain beach in Longwangmiao Formation, Moxi area, Sichuan Basin were high yield with pores in millimeter or centimeter level, which are considered as the product of hydrothermal dissolution. While the grain beach dolomite reservoirs without the

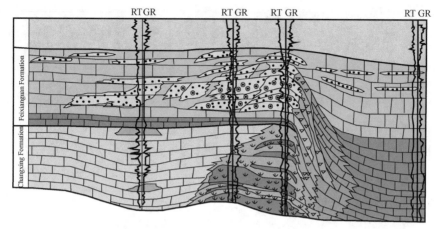

(a) Development model of reef and beach in platform margin

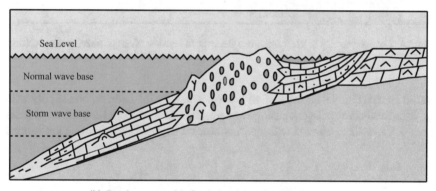

(b) Development model of grain-beach in carbonate gentle slope

Fig. 4.2 Sedimentary models two types of of carbonate reef-beach

hydrothermal reformation are developed with much lower production and strong reservoir heterogeneity as the lack of pores in millimeter or centimeter level, even with the development of acicular matrix pores.

4.1.3 Effectiveness of Passage System and Scale of Transmission and Distribution

The communication of passage system is very important as the source rock and reservoir of carbonate strata are usually not directly contacted. The carbonate passage system is mainly composed of faults, fractures and unconformities, and plays an important role in the communication on strata with the strong heterogeneity in carbonate reservoirs. The Ordovician oil and gas in Tabei and Tazhong of the

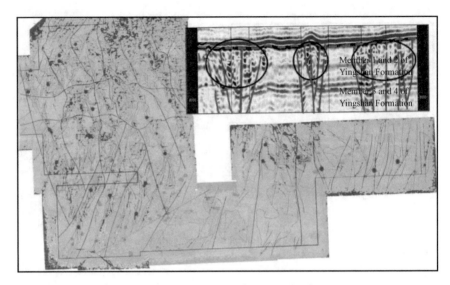

Fig. 4.3 Change rate map of RMS amplitude along the top surface of carbonate reservoir (0–30 ms) in Halahatang Oilfield

Tarim Basin (Fig. 4.4) accumulated in large scale, by the hydrocarbon supply with fault-unconformity network passage system, and transported and controlled by main faults. A oversized oil area with multi-strata and large-area distribution are formed

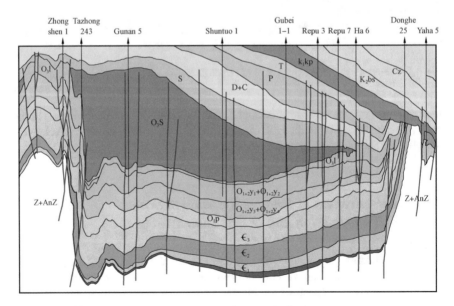

Fig. 4.4 Accumulation model of Tazhong and Tabei in Tarim Basin

by aujusted with unconformity, efficient passage system and hydrocarbon kitchen connected with faults. Exploration practice shows that 85% of the high-yield wells are distributed within 1000 m from the main faults.

The reef-beach gas reservoirs in the Sichuan Basin were developed with different accumulation combinations in different positions. The reef-beach gas reservoirs with multi-strata are developed in the platform margin of east of the trough, due to the connection of multiple sets of source rocks by large faults and vertical transportation of the major faults. The fullness of gas reservoir is 87%–94%, and the reserve abundance is $(5.3–78) \times 10^8$ m^3/km^2, with an average of 18.6×10^8 m^3/km^2. The reef-beach gas reservoirs in the platform margin of west of the trough, were developed with the uneven transportation by faults and fractures. The fullness of the gas reservoirs is 58.2%–77.8%, and the reserve abundance is $(2.3–5.4) \times 10^8$ m^3/km^2, with an average of 3.3×10^8 m^3/km^2. The reef-beach gas reservoirs inside platform are mainly transported by small faults and fractures. The fullness of gas reservoirs is 36.3%, obtained by only a few wells, and the reservoir forming efficiency is low.

The transportaion ways in different passage systems are developed by the different types of source reservoir collocation, and mainly in the following forms.

(1)   Lateral transportation by large-scale unconformities in the combination of "source-reservoir integration"

Source and reservoir were directly contacted in this kind of source-reservoir configuration, including the types of "sandwich" and "pizza". The former refers to the interbedding distribution of multilayers source and reservoir, while the latter refers to the direct contact of source and reservoir. The weathering crust of the Leikoupo Formation in Sichuan Basin was directly covered by the coal-measure source rocks of Xujiahe Formation. Oil and gas accumulation was controlled by the weathering crust reservoirs, and the source-reservoir connection was developed by the passage system mainly composed by unconformities. As a result, the large-scale accumulation can be developed of lateral source rock-reservoir collocation, which was formed by high-quality source rocks and large-sclae reservoirs linked with the unconformities (Fig. 4.5).

(2)   Transportation by faults and fractures in the combination of "lifting pattern"

In the Tabei area of Tarim Basin, the oil–gas transportation of "lifting pattern" was developed as the oil and gas was migrated upward by oil–water buoyancy along the fracture-cave system of carbonate rock with the lateral hydrocarbon supply of source rock in depression (Fig. 4.6). This kind of oil–gas migration model was mainly developed in the ancient uplift area of the basin which was evolved with inheritance, and the migration mechanism can be summarized as " energy stored by buoyancy, accumulated in the caves, transported by fractures, migrated and accumulated, accumulated stratoid ".

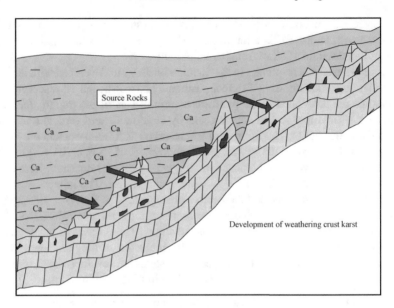

Fig. 4.5 Accumulation model of unconformity transportation in Leikoupo Formation of Sichuan Basin

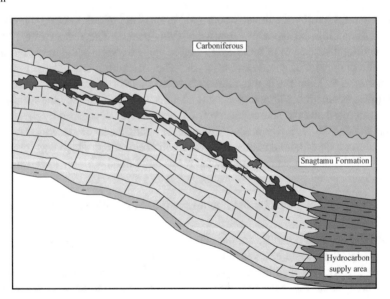

Fig. 4.6 Transportation and accumulation model of fractures and caves in Yingshan Formation of Ordovician in Tarim Basin

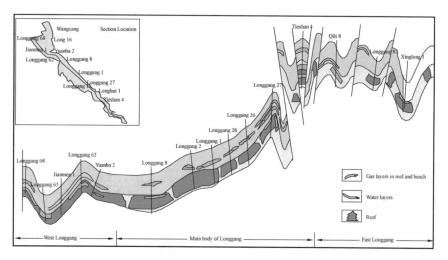

Fig. 4.7 Oil and gas migration and accumulation model of "shift and joint" from Changxing Formation to Feixianguan Formation in Sichuan Basin

(3)   Transportation by faults and fractures in the combination of "shift and joint"

The source was developed under the flank of reservoir, and not direct contact with the reservoir or limited contact. The oil and gas was accumulated to the reservoirs through the transportation media such as fault and unconformity, since it was oil and gas are generated from the source rocks (Fig. 4.7). The oil and gas are transported by faults from Changxing Formation to Feixianguan Formation in Sichuan Basin, and the oil and gas in the lower part is moved upwards with distribution of "beaded" pattern along the edge of the platform The complex characteristic of oil and gas accumulation as "one reef, one beach and one reservoir" was formed by the difference in the configuration of faults and heterogeneous reservoirs.

4.1.4   Sealing and Effectiveness of Caprocks

(1)   Good sealing property of deep gypsum-salt rock

Gypsum-salt rock is brittle and difficult to be an effective cap rock under the condition of shallow buried depth as well as low temperature and pressure, as the brittle plasticity of gypsum-salt rock is mainly related to temperature and pressure. And the gypsum rocks can be developed as effective caprocks with the transformation from brittle to plastic under the condition of high temperature and high pressure.

The physical characteristics of gypsum-bearing dolomites under different temperature and pressure were revealed through the simulation of brittle-plastic property in gypsum-bearing dolomite samples from different areas under 8 groups of different

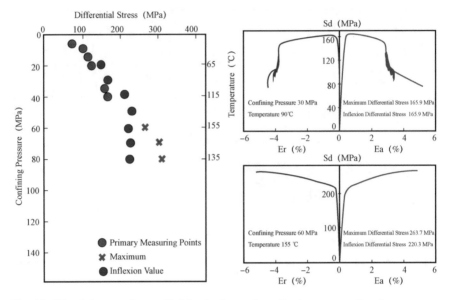

Fig. 4.8 Stimulation experiment of brittle-plastic transformation in gypsum-salt rocks

temperature and pressure. The results show that (Fig. 4.8), the gypsum-bearing dolomites are not always developed with plastic property but with a conversion process: the gypsum rocks are still developed as brittle property with confining pressure less than 50 Mpa and the temperature less than 135 °C; and the gypsum rocks are developed as plastic rocks with confining pressure more than 60 Mpa and the temperature more than 155 °C; between them, they show transition state from brittle to plastic. The matching results with the under-salt burial history of actual wells in Cambrian of Sichuan Basin (Fig. 4.9) show that, the gypsum-salt rocks were still developed with brittle property and poor sealing in the early stage of oil generation; the gypsum-salt rocks were developed with transition state from brittle to plastic and had limited sealing property in the peak period of oil generation;the gypsum-salt rocks were developed with plastic property in the period of hydrocarbon cracking to gas, with effective sealing of under-salt oil and gas. Therefore, the match between oil generation period of source rock and brittle-plastic transition period of gypsum-salt rock is beneficial to the preservation of under-salt oil and gas. For example, during the main accumulation period of Early Cretaceous in the central of Ordos Basin with temperature of 130–160 °C, the gypsum-salt rocks had been transformed into plasticity and developed with good sealing property. As a result, the under-salt exploration is worthy of expecting.

By the structural deformation simulation, two types of traps and good migration pathway were developed with the uncoordinated deformation of up-salt and under-salt under tectonic compression environment. Physical and mathematical simulation experiments (Fig. 4.10) show that, the up-salt and under-salt traps can be formed by the uncoordinated deformation of up-salt and under-salt strata caused by the

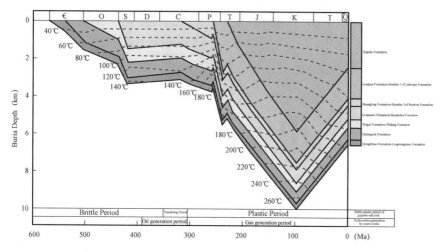

Fig. 4.9 Matching relationship between brittle-plastic transformation and evolution history of hydrocarbon generation in gypsum-salt rocks

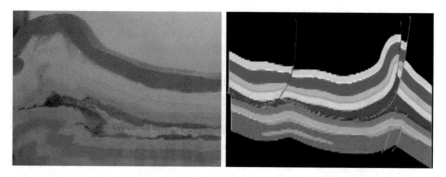

Fig. 4.10 Stimulation experiment of brittle-plastic transformation in gypsum-salt rocks

detachment of gypsum-salt rocks. The sealing continuity of gypsum-salt rock was destroyed by fault and salt misplacement, while the gypsum-salt rock can still be the main migration path way of oil and gas; In the areas where the source and reservoir are well configured, the structural traps related to salt are expected to become new exploration fields.

(2)   Controlling factors of sealing property in high-quality caprocks

High-quality preservation conditions are indispensable for oil and gas accumulation, especially for the ancient marine carbonate rocks undergoing multi-stage tectonic movements, as well as unavoidable destruction and adjustment in oil and gas reservoirs.Statistics show that the distribution of large oil and gas fields in China was controlled by caprocks, and three types of caprocks, such as evaporite, mudstone and shale, can be as high-quality caprocks (Fig. 4.11). Recent exploration has proved that

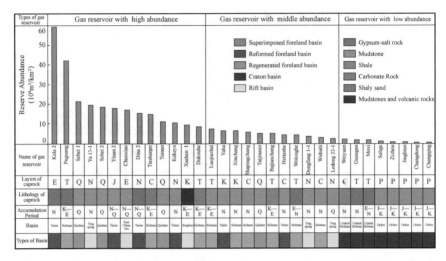

Fig. 4.11 Accumulation period and caprocks distribution stara of typical large gas fields in China

the oil and gas accumulation in deep marine carbonate reservoir must be developed with good sealing conditions. For example, three sets of major strata in oversized Anyue Gasfield are closely associated with the direct caprocks, which is benefited from the effective sealing of high-quality caprocks composed of shale and gypsum-salt rock. The carbonate reservoirs found in Gaoshi 1, Moxi 8 and Nvji1 in oversized Anyue Gasfield of Sichuan (Fig. 4.12) are developed with the three sets of direct caprocks: the first set is mainly composed of marlstone and gypsum-salt rock in Gaotai Formation, with thickness of 40–70 m, pressure in underlying gas reservoir of 12–65 Mpa, thickness in gas reservoir of 1.56–1.65 m, proved reserves of 4403 $\times$ 10^8 m^3; the second set is mainly composed of argillaceous rock in Qiongzhusi Formation, with thickness of 80–150 m, pressure in underlying gas reservoir of 40–150 Mpa, thickness in gas reservoir of 1.12–1.13 m, proved reserves of 2200 $\times$ 10^8

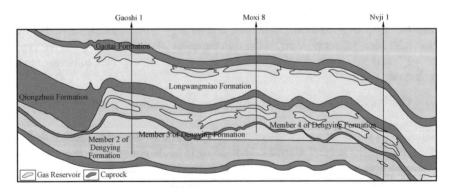

Fig. 4.12 Combination of major productive strata and direct caprocks in Anyue Gasfield, Sichuan

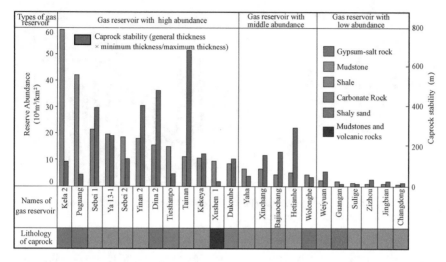

Fig. 4.13 Relationship between reserves abundance grade and stability of caprocks in typical large gas fields

m³; the third set is mainly composed of argillaceous rock in Member 3 of Dengying Formation, with thickness of 10–35 m, pressure in underlying gas reservoir of 15–110 Mpa, thickness in gas reservoir of 1.10 m, and controlled reserves of 2300 × 10⁸ m³. At the same time, the reserve abundance of gas reservoir is closely related to the stability of caprocks (Fig. 4.13). The more stable the caprocks distribution is, the stronger the preservation capacity is.

In addition, the displacement pressure of caprocks is also an important parameter reflecting its control strength. The abnormal high pressure is often developed in gas reservoirs with medium and high abundance, and abnormal low pressure is often developed in gas reservoirs with low abundance. The pressure system is closely related to the displacement pressure of caprocks (Fig. 4.14).

4.2 Accumulation Mechanism of Deep Oil and Gas in Cross-Tectonic Period

The connotation of cross-tectonic accumulation is the coupling of "progressive buying" and "annealing heating". The source rocks had been in the liquid window for a long time, and had not been damaged by multi-phase tectonic movements, so the liquid hydrocarbon reserves exceed previous geological understanding and the question of long-period storage and late-stage accumulation of liquid hydrocarbon in marine carbonate rocks (Chengzao et al. 2006, 2007) is also answered. For the understanding of accumulation of cross-tectonic periods, two kinds of mechanisms were proposed. One is that source rocks were in the liquid window for a long period,

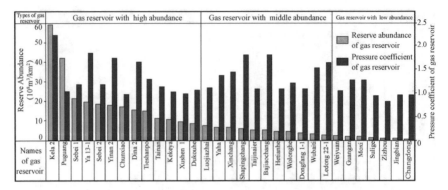

Fig. 4.14 The relationship between the abundance grade of reserves and the pressure coefficient of gas reservoirs in typical large gas fields

and the structural damage was avoided to the maximum extent; the other is the cross-tectonic gas reservoir with the accumulation and effective preservation of large-scale natural gas in deep ancient strata, caused by the multi-factor superposition of transformation in hydrocarbon phases, late-stage gas generation of multi-source kitchens and inherited structural preservation.

4.2.1   Oil Accumulation in Cross-Tectonic Period

(1)   Thermal stability of crude oil

Two meanings are included in the concept of thermal stability of crude oil: one is the disappearance temperature of crude oil with separate phase, that is the maximum storage temperature of pure reservoir; the other is the complete disappearance temperature of crude oil, that is the liquid movable hydrocarbon is basically disappeared. Claypool et al. (1989) proposed that the crude oil with separate phase would disappear when the crude oil conversion rate reached 62.5%. However, it is also believed that when the crude oil conversion reaches about 51%, the disappearance of crude oil with separate phase will happen. The results show that there is little difference between the storage temperatures of the two types of crude oil under geological conditions. In this simulation research, 62.5% was used. The simulation experiment of crude oil cracking is shown in Fig. 4.15. The low limit of crude oil preservation can reach more than 9 km with the disappearance temperature in separate phase of 200 °C, under the pressure of 100 MPa and general burial conditions (2 °C/Ma, 20 °C/km). The preservation temperature of reservoir with pressure of 150 MPa is nearly 30 °C higher than that of reservoir with pressure of 50 MPa, and the preservation depth moved down about 2.5 km. It can be seen that the high pressure should be paid enough attention in deep oil and gas resource evaluation, exploration and development due to the significant impact on the thermal stability of crude oil.

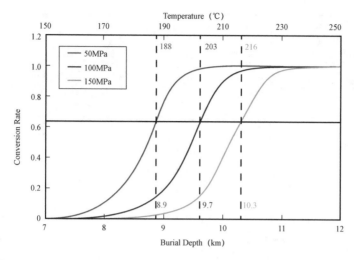

Fig. 4.15 Geological deduction of crude oil cracking under different pressures (2 °C/Ma, 20 °C/km)

(2)   The delay of hydrocarbon generation with the coupling of "progressive buring" and "annealing heating"

A large amount of exploratory researches on the hydrocarbon generation and accumulation of the ancient marine strata have been carried. The general view is developed as the source rocks of ancient marine strata (Cambrian-Ordovician) in China were matured in the Late Caledonian-Early Hercynian, and now in the high-over mature stage. The scale and exploration potential of the original oil and gas reservoir were relatively limited as the reservoirs formed in the early period were largely damaged and transformed in the geological history. As a result, there are many doubts about the exploration of carbonate strata, especially the exploration potential. However, from the perspective of thermal evolution history, the marine source rocks was basically in the annealing process, with the thermal evolution history of high geo-temperature in early stage and low geo-temperature in late stage (gradually cooled down since Mesozoic) in petroliferous basin in central and western China. Moreover, a large number of simulation experiments in hydrocarbon generation which are close to the real underground environment (temperature–pressure co-control) show that, the Ro of oil generation peak is 1.5% and Ro of the gas generation peak is more than 1.8%, under the thermal history of gradual decreased geothermal gradient, geological conditions of deposition-denudation repeats and rapid-deep burial in the late stage. Compared with Tissot's hydrocarbon generation results of mature kerogen, this result shows the characteristics of obvious delay in hydrocarbon generation. That is to say, the generation and expulsion of hydrocarbon in some ancient source rocks can be delayed under the coupling of "progressive burying" and "annealing heating", so the generation and accumulation of large amount of liquid hydrocarbon can still be developed in the late stage.

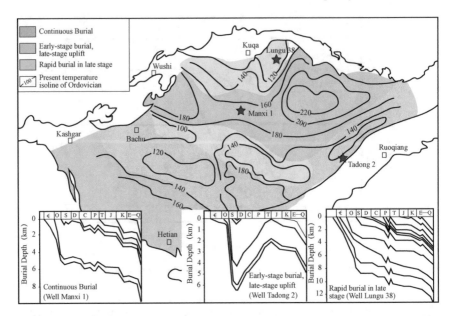

Fig. 4.16 Three burial evolution models and distribution map of Cambrian-Ordovician in Tarim Basin

The ancient marine source rocks of Cambrian-Ordovician in Tarim Basin were basically in the annealing process, after the thermal evolution history of high geo-temperature in early stage and low geo-temperature in late stage (gradually cooled down since Mesozoic). The history of burial and evolution in different structural areas were obviously different due to the obvious differential subsidence among the basins. Three types of burial evolution models (Fig. 4.16) were developed in the Cambrian-Ordovician of the basin, such as the model of continuous progressive burial represented by Manxi 1, the model of early-stage deep burial and late-stage uplift represented by Tadong 2, and the model of early-stage continuous shallow burial and late-stage rapid burial represented by Lungu 38. In the third model, the ancient source rocks stayed in the "liquid window" for 400 million years, from late Silurian to Paleogene-Neogene, with the combination between burial process and "annealing" process in geothermal field. As a result, some ancient source rocks stayed in the "liquid window" for a long time and the oil and gas can avoid being destroyed by multi-stage tectonic movements, and the generation and accumulation of large amount of liquid hydrocarbon can be developed since Paleogene.

(3)   Oil accumulation of cross-tectonic periods in Tarim Basin

It is found that liquid hydrocarbons were still developed below the buried depth of 7000 m in Tarim Basin. In the coupling effect of "annealing" geothermal field and progressive burial (Fig. 4.17), some of the ancient source rocks were stayed in the generation range of liquid hydrocarbon for a long time, and the liquid window

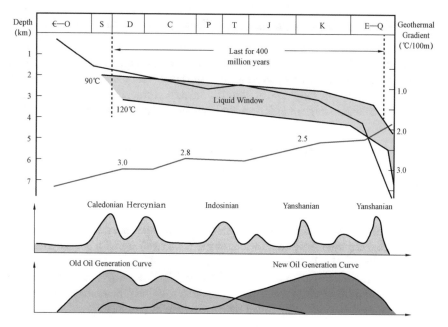

Fig. 4.17 Accumulation in across-tectonic periods with the coupling of "progressive burying" and "annealing heating"

can last for as long as 400 million years. As a result, the liquid hydrocarbons can be preserved with the multi-period destruction of major tectonic movements. The reservoir accumulation shows that, the oil and gas reservoirs found in the platform area of Tarim Basin were mainly formed in the late stage, only 2–5 Ma from now.

The potential of petroleum resources in carbonate strata of Tarim Basin was reevaluated under the guidance of the study in late-stage hydrocarbon generation and accumulation of the ancient source rocks across the major tectonic periods. The area of source rocks which have been in the range of "liquid window" for a long time (the source rocks developed with the burial evolution model represented by Lungu 38) is about 15×10^4 km^2, accounting for 58% of the area of marine source rocks. The amount of petroleum geological resources has increased from 42×10^8 t to 85×10^8 t, with a net increase of 43×10^8 t, twice as much as the previous evaluation, indicating that considerable oil resources are still retained in ancient carbonate rocks of the platform area in Tarim Basin. In this way, the exploration potential of carbonate in the platform area is greatly enhanced, as well as the confidence of deep oil exploration.

4.2.2   Gas Accumulation in Cross-Tectonic Periods

The key factors for the large-scale accumulation in cross-tectonic periods of natural gas are superposition of many factors such as the preservation and transformation of hydrocarbon phase, the long time of liquid hydrocarbon cracking, the continuous gas generation in the late stage of multi-source kitchens, and the preservation of inherited structures.

(1)   The preservation and transformation of hydrocarbon phases are important ways of accumulation in cross-tectonic periods

①   Thermal stability of natural gas

Through the simulation experiment, it was found that the pressure has certain influence on the thermodynamic equilibrium of methane cracking with the range of 10–1000 MPa (Fig. 4.18). The equilibrium concentration of methane cracking products is obviously reduced by the increase of pressure, which generally does not change the shape of the equilibrium curve, but moves it to a higher temperature. That is, the cracking of methane can be inhibited by the increase of pressure.

②   The preservation of hydrocarbon phases

The kinetic parameters of generation and cracking of methane, ethane, propane, butane/pentane can be calculated by the yields of gaseous hydrocarbon products of crude oil cracking at different thermal maturities (Fig. 4.19). The simulated values of kinetics can be fitted well with the experimental results, indicating that these kinetic parameters have high reliability. It can be seen that ethane has the highest thermal stability, followed by propane, and butane/pentane is the lowest. Based on these kinetic parameters, it can be predicted that, under general burial conditions (Fig. 4.20), the cracking of butane/pentane will begin at 200 °C and burial depth of about 9 km, while the cracking completely of propane was 210 °C and burial depth

Fig. 4.18 Conversion rate of CH_4 at different pressures in thermodynamic equilibrium conditions

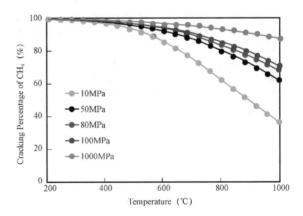

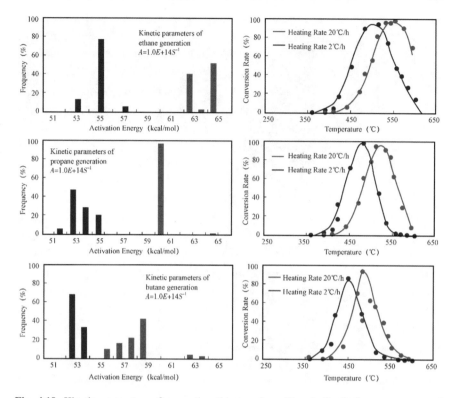

Fig. 4.19 Kinetic parameters of generation (blue) and cracking (red) of ethane, propane, and butane/pentane and the fitting curve between simulation values and experimental results

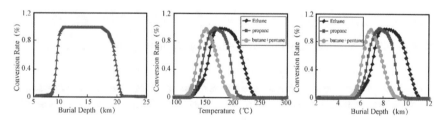

Fig. 4.20 Evolution law of methane, ethane, propane, butane/pentane under the difference of temperature and burial depth

of 10 km, and ethane has the highest thermal stability and can be preserved to 250 °C and 11 km. Similarly, the cracking curve of methane can be calculated by assuming a geothermal gradient of 2 °C/Ma (Fig. 4.20). It is predicted that the cracking of methane will not happen until the temperature is as high as 1000 °C, and the burial depth of about 20 km.

Based on above, the lower limit of depth and corresponding thermal maturity of crude oil and various gaseous hydrocarbons are summarized (Fig. 4.21). In the early

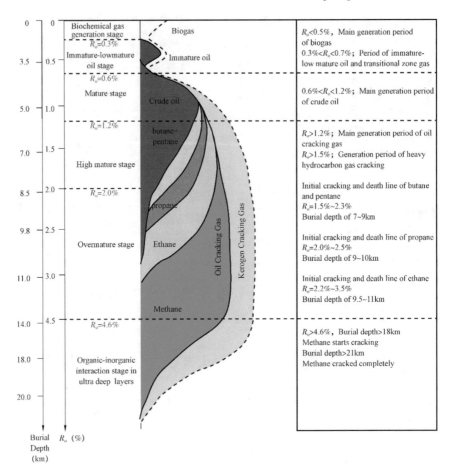

Fig. 4.21 Buried depth and lower limit of Easy Ro of crude oil and gaseous hydrocarbon

stage of thermal evolution in organic matters, the state of crude oil is relatively stable, as the enough potential energy cannot be provided by thermal stress in the environment for the cracking reaction of each component in the crude oil. Since the later stage of maturity (Easy Ro > 1.0%), the cracking of crude oil gradually enters the wet-gas stage as it gradually undergone thermal cracking under increasing thermal stress, such as the C–C bond fracture and formation of short-chain aliphatic hydrocarbon in the saturated alkyl, and the demethylating reaction of aromatic hydrocarbon and pyrobitumen. At this time, the production rate of methane is low, mainly because the high activation energy for generating methane is high and cannot be carried out smoothly. The secondary cracking is developed in generated heavy hydrocarbon gas with the continues increase of cracking rate in crude oil caused by the increase of thermal stress, that is, the fracture of C=C bond and the generation of methane in aliphatic chains of $C_{2 \sim 5}$. The crude oil is almost completely cracked when the Easy

Ro is more than 2.5% and the buried depth is about 9 km. At this time, in addition to methane generated from crude oil cracking, butane/pentane and propane disappear in turn with the large amount of secondary cracking in heavy hydrocarbon gas. Finally, ethane is also completely converted into methane with the Easy Ro of more than 3.5% and the buried depth of about 11 km. Among all kinds of cracking products in crude oil, methane has the highest thermal stability, which cannot be cracked until the burial depth is 18 km in general geological condition (2 °C/Ma, 20 °C/km), and it can be preserved to a depth greater than 20 km.

Under general geological condition (2 °C/Ma, 20 °C/km), the retention depths of methane, ethane, propanoid and butyl/pentylene in natural gas are 20 km, 11 km, 10 km and 9 km respectively, which shows that, natural gas can be preserved for a long time under the condition of deep high pressure, and crude oil cracking gas and heavy hydrocarbon gas further cracking gas can last up to nearly ten thousand meters in depth, which is the internal mechanism and material guarantee of gas accumulation in the cross-tectonic periods.

③ Phases transition of hydrocarbons

Three stages of gas accumulation are developed in deep strata of China, such as the formation of paleo oil reservoir, the formation of paleo gas reservoir by oil cracking, and the formation of present gas reservoir by late adjustment (Fig. 4.22). The large gas field can be formed by crossing the "liquid window" and entering the "gas window" when developing into the stage of high-over mature due to the continuous deep burial. The formation of paleo oil reservoirs (liquid hydrocarbon inclusions, with homogenization temperature of 100–160 °C) and the formation of oil-cracking

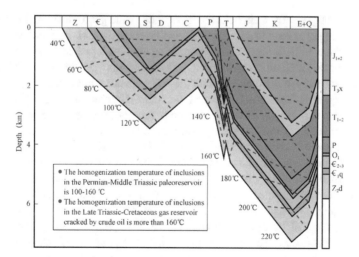

Fig. 4.22 Thermal evolution of source rocks of Sinian and hydrocarbon filling history in Gaoshi 1

gas reservoirs (gas–liquid two-phase inclusions, with homogenization temperature of more than 160 °C) were included in the hydrocarbon filling events revealed by inclusions.

(2) Long time of liquid hydrocarbon cracking is favorable for oil and gas preservation and hydrocarbon supply in late stage

The crackings of liquid hydrocarbon were developed with long time both in the Sichuan Basin with high geothermal gradient and Tarim Basin with low geothermal gradient (Hui et al. 2006; Wenzhi et al. 2011; Zhang Shuichang et al. 2011), and are advantageous to the continuous hydrocarbon supply in the late stage and the accumulation crossing the tectonic periods. The deep marine strata in Sichuan Basin are dominated by gas, and even C_{2+} gas craked in some areas, with long time of cracking and all-time gas generation (Fig. 2–6, Shuichang et al. 2013). The cracking of crude oil in Tarim Basin is insufficient, and the complete cracking has later and longer time limit. The simulation experiments showed that the temperature of condensate cracking in South Sichuan can reach 240 °C which is also the evidence of long cracking time in liquid hydrocarbon.

In addition, previous studies have pointed out that methane is the most stable hydrocarbon. Under catalytic conditions, the methane can be preserved in the temperature of 700 °C, and can be preserved in the temperature of more than 1200 °C without catalyst. It also shows it has the material basis of late-stage continuous hydrocarbon supply in deep strata crossing tectonic periods.

(3) Continuous gas generation in late stage of multi-source kitchens provided material source for the accumulation crossing tectonic periods

Many types of gas source kitchens are developed in the geological conditions, such as the cracking of kerogen and the cracking of liquid hydrocarbon. More types of gas source kitchens in liquid hydrocarbon cracking are developed, such as the cracking in the aggregate paleo oil reservoir, half-way cracking in "pan-reservoir" with semi-accumulation and semi-dispersion, and late-stage cracking in remained hydrocarbon kitchens (Fig. 2–4). The material source was provided by continuous gas generation in late stage of multiple source kitchens for the accumulation crossing tectonic periods. The evidence of hydrocarbon supply from multiple source kitchens was preserved in Gaoshiti-Moxi Gasfield of Sichuan Basin. The boundary of paleo oil reservoir is outlined by the contour of asphalt content 0.5% and the development of rounded asphalt, which was mainly formed in the high part of paleo uplift and steep slope zone. On the gentle slope, the asphalt content is low, and the liquid hydrocarbon concentration is not high. There is a phenomenon of concentration locally, showing a distribution type of "semi-accumulation and semi-dispersion".Different occurrence states of reservoir bitumen and relationship with rock minerals formation reflected multi-stage filling and late cracking of hydrocarbons, including the following four modles (Fig. 4.23). Two-stage asphalt rings formed by two-stage hydrocarbon filling and late degradation; asphalt grid pores formed by early filling and late cracking; two-stage asphalt distribution formed by two-stage hydrocarbon filling and precipitation

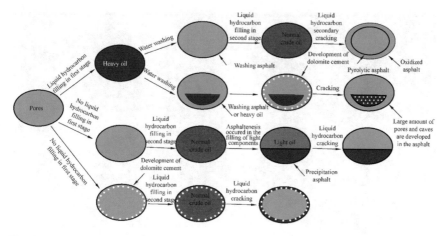

Fig. 4.23 The filling and cracking of liquid hydrocarbon reflected by reservoir bitumen in multi-occurrence states

of heavy components, with up and down; and asphalt with single ring-belt distribution formed by early-stage filling and late-stage cracking, The development of tar pitch in reservoirs with various occurrence states is a direct sign of continuous hydrocarbon supply from various hydrocarbon source kitchens.

(4)   The preservation of inherited structures is the key factor for accumulation crossing tectonic periods

The late preservation of large oil and gas fields after reservoir formation is of great importance with the characteristics of multi-cycle development in China superimposed basins. Taking the formation of Gaoshiti-Moxi Gasfield in Sichuan Basin as an example, one of the key factors is that the region lied in the development area of inherited paleouplift for a long time, which is the favorable orientation for oil and gas. The paleouplift in Gaoshiti-Moxi area was developed with inheritance and strong stability (Fig. 4.24), and lied in Ziyang area high part of paleouplift in early stage of, and in slope belt in Himalayan, although the superposition of multi-stage structures, and the migration of paleouplift axis from north to south during Tongwan Period (Zecheng et al. 2014) to Himalayan. Weiyuan area lied in the slope zone in early stage of, and structural high part in Himalayan. However, the oil and gas accumulation was seriously damaged and transformed by multi-stage tectonic activities, expecially Himalayan movement. The large-scale gas fields were formed in Gaoshiti-Moxi area with less effect on tectonic movements, and completely destroyed at basin margin with serious destruction of tectonic movements.

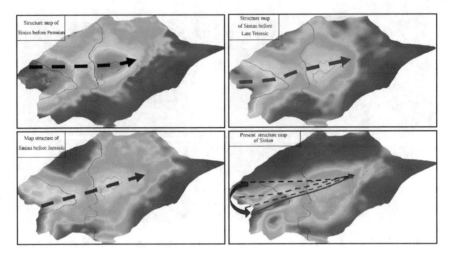

Fig. 4.24 Maps of Paleotectonic evolution and migration of paleouplift axial in top Sinian of Sichuan Basin

4.2.3 Gas Accumulation Crossing Tectonic Periods in the Complex Tectonic Belts Western of Sichuan

The gas accumulation crossing tectonic periods was developed in slope of paleouplift of Gaoshiti-Moxi area in the central Sichuan Basin, with major controlling factors of relatively stable structure and little effect on late-stage transformation. However, for the research of gas accumulation crossing tectonic periods, another important problem is the trap completeness and accumulation effectiveness after multi-stage structural superposition in complex structural areas.

(1)   Analysis of structural modeling and evolution of northern Longmenshan

The structural style of each submember in the northern Longmenshan is different, with the characteristics of multi-stage and layered detachment structure and deformation. Four different types of structural systems formed in different structural strata and periods can be distinguished: (1) deformation system in Presinian strata: the thrust fault-block based on detachment layers composed of ductile shear zone in basement, and mainly activity in Cenozoic. It is worth noting that the structural wedges superimposed on basement were mainly developed as front expansion in north submember, and back expansion in middle and south submembers. (2) The imbricate structure, superimposed structural wedges and related fold systems in Indosinian: the area restricted by the step thrust-fault and the unconformity between Member 4 of Xujiahe Formation and Baitianba Formation, and the major deformation occurred in the middle structural layer. The structural system formed by the local main unconformities had great contribution to the structural shortening of Longmenshan by controlling the fault transition and fault-propagation fold in the front of the nappe belt, the local wedge structure and imbricate structure. (3) The step fault-bend fold and structural emplacement system in Yanshanian: the main deformation occurred in the middle-shallow strata, characterized by the involved structure deformation in Jurassic. The sheer fault-bend fold was controlled by structural wedge mainly composed by large-scale wide-gentle fault-bend fold and local back-thrust faults with the fault emplacement developed up and toward the basin in Yanshanian. The imbricated structures are not obviously developed in Indosinian structures area, and the structures produced in the Jurassic are not completely preserved due to late-stage transformation and destruction. (4) Activated-thrust structure and wedge structure in Himalayan: with the relatively large fault throw of activated fault in Himalayan, the marine strata such as Paleozoic strata and Middle-Lower Triassic strata were directly thrusted above the upper Jurassic strata, and showing strong destruction and transformation to the pre-existing structures in Indosinian and Yanshanian. Taking Xiangshui Fault as an example, it has the biggest fault throw between Xiangshui and Anxian. The anticline of Yanshanian was only completely developed in fault foot-wall, and damaged to varying degrees in hanging wall. The fault extended northward to the Haitangpu anticline, with gradually decreased fault throw.

The evolution models of foreland basins in northern Longmenshan and northeast Sichuan were preliminarily established, and the key transformation periods of 5–6 stages were determined, combined with the research of predecessors (Zhengwu et al. 1996; Chuanbo et al. 2007; Richardson et al. 2008) on the geotectonic background and tectonic-sedimentary paleogeography of research area, based on the stratigraphic and structural analysis of the northern Longmenshan and northwest Sichuan. (1) Before the Late Triassic, the crust of Longmenshan area had been in tension state for a long time. The rift trough, passive continental margin and many tensional fractures developed in west margin of Yangtze Block in Caledonian were formed by "Emei Tafrogeny" from Permian to Triassic. And these tensional fractures became the boundary of Longmenshan formation. (2) early Late Triassic, the Indosinian movement in Longmenshan area was characterized by micro-angle unconformity and parallel unconformity between Maantang Formation and Xiaotangzi Formation in Late Triassic, as well as Leikoupo Formation and Tianjingshan Formation in Middle Triassic. (3) The II episode of Indosinian Movement developed at the end of Member 3 in Xujiahe Formation was mainly characterized by micro-angular unconformity and parallel unconformity between Member 3 and 4 of Xujiahe Formation (Fig. 4.25). In middle Indosinian, the tectonic movement in northern part of Longmenshan was mainly developed as uplift, with weak folding. (4) The tectonic movement in Late Indosinian was occurred in the end of Xujiahe Formation (Member 5) in Late Triassic. This tectonic movement is the strongest in the Indosinian movement of Longmenshan area, with the distribution range of whole Longmenshan area, and the strongest area in the northern part of Longmenshan area. The geological phenomenon is mainly developed as angular unconformity between Baitianba Formation and Xujiahe Formation in Jurassic (Fig. 4.26). The properties of this movement are strong fold and thrust nappe, which determine the attribute of thrust-nappe structure in Longmenshan area. (5) Himalayan movement was the latest tectonic movement in Longmenshan area, as well as another key period in the evolution

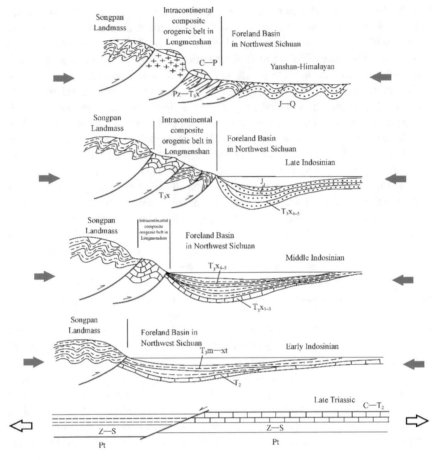

Fig. 4.25 The model of formation and evolution of the north Longmenshan area and the foreland basin in northwest Sichuan

history of Longmenshan area, western Sichuan. Sichuan Basin not only ended the continental sedimentary history,but deformed the strata forming after Jurassic and revived the thrust-nappe structure in Indosinian. The northern part of Longmenshan area is mainly characterized by nappe thrust in shallow strata and structural wedge in middle-deep strata.

(2)   The control of structural evolution on oil and gas in the northern part of Longmenshan area

The degree of exploration is still low even with the transportation of oil and gas exploration to the northwest of the Sichuan Basin, which is mainly caused by the multi-periods of trusts and complex structures. Combined with the structural deformation characteristics of the northern Longmenshan in Sichuan Basin, the petroliferous properties are analyzed.

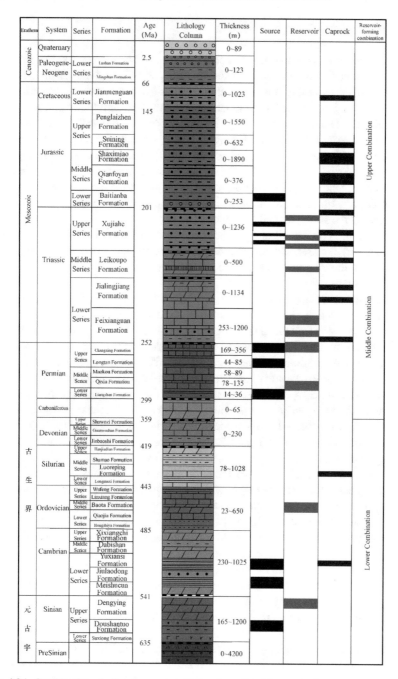

Fig. 4.26 Combination of oil and gas accumulation in northern Longmenshan area

① Source-reservoir-cap combination

The hydrocarbon accumulation is affected by the back mountain belt, the front mountain belt and the fore mountain belt in the northern part of Longmenshan area. Among them, the hydrocarbon is not accumulated in the back mountain belt, while accumulated as Jinzishan-Qinglinkou-Houba hydrocarbon accumulation belt in the fore mountain belt. The front mountain belt is characterized by the formation of short-axis anticlines, such as Tianjingshan, Kuangshanliang, Nianziba, Yangquan, Wuhuashan and so on, with the weak thrust nappe in belt of Kuangshanliang-Tianjingshan-Erlangmiao thrust anticlines. The Cambrian paleo oil reservoir is formed in those anticlines with the secondary migration of oil and gas by the faults, which are generated in Doushantuo Formation of Sinian. And the Cambrian paleo oil reservoir is destroyed by the strong tectonic actions in late stage. At the same time, the oil reservoirs are formed by the oil and gas migrated from paleo oil reservoirs to strata in Devonian and Jurassic, through the faults connecting the strata of Devonian and Jurassic, and the oil and gas also can be formed by crude oil exposed on the surface in research area.

The nappe movement is well matched with the migration of oil and gas in Longmenshan area, that is, the formation time of surface structural traps is earlier than the migration time of oil and gas. Longmenshan orogeny began in the Late Triassic. The high-quality traps for the later oil and gas accumulation were developed under the strong northwest compressive force, such as the Tianjingshan and Kuangshanliang surface traps. In the early Jurassic, the source rocks in the deep strata began to mature. Under the compressive force of nappe action, the oil and gas in the source rocks migrated to the surface along the deep-large faults, on the other hand, oil and gas reservoirs were formed with the accumulation in early-stage surface traps during the lateral migration of oil and gas. The oil and gas reservoirs formed in early stage are easy to be destroyed in the late-stage structural superposition, as the oil and gas are migrated to the surface by the faults developed in the traps under the continuous compression in the late stage, which are caused by the continuous orogeny in Longmenshan area. The surface geological survey shows that, the traps with oil and gas accumulation formed in early stage are destroyed in the later stage of structural superposition, which is indicated by the large amount of asphalt remained on the surface structures such as Kuangshanliang and Nianziba. At the same time, the repeated lost in Feixianguan Formation can be confirmed by the Kuang 2, as a result, the heavy fracture of rocks near detachment layers can be speculated, as well as the damage of sealing property in the caprocks. Howerer, it can be confirmed that the reservoirs have been damaged by the fresh water samples although encountering a high-quality reservoir.

The source-reservoir-cap combinations can be divided into upper, middle and lower combinations through the structural stratification characteristics and vertical distribution characteristics in north Longmenshan area (Fig. 4.26). The upper combination is developed in the continental strata above the Leikoupo Formation of Middle Triassic. The source rocks are mainly composed of dark mudstones in Baitianba Formation and Xujiahe Formation, with a cumulative thickness of more than 300 m and the total organic carbon content of 0.4%–1.1%, and the peak period of oil generation in the early Cretaceous. The reservoirs are mainly developed as sandstones in Member 2 and 4 of Xujiahe Formation, which are the sedimentary system of delta-river, with an average porosity of 5.98%, an average permeability of 2.84 mD. The caprocks are main thick mudstones in Middle Jurassic and mudstone interbeds in Xujiahe Formation. The middle combination is developed in marine strata between Permian and Leikoupo Formation of Middle Triassic. The source rocks are mainly composed of carbonate rocks, marls and coal-bearing formations in Permian, with a cumulative thickness of about 250 m and a total organic carbon content of 0.4%–3.72%, high-over mature in organic matters and the peak of oil generation in Late Jurassic. The reservoirs are mainly composed of dolomite and limestone in Qixia Formation, with porosity of 1.3%–5.16%, permeability of 0.47–567 mD, and a strong heterogeneous in physical properties. The oolitic limestones of Member 3 in Feixianguan Formation have porosity of 3%–7.32%, the average porosity of dolomitized organic reef beach in Changxing Formation is 6.5%. The caprocks are mainly composed of gypsum-salt rocks in Leikoupo Formation of Middle Triassic-Jialingjiang Formation in Lower Triassic. The large-scale gas fields are most likely to be found in middle combination. The lower combination is developed in the deep marine strata. At present, the source-reservoir-cap combination is potentially developed and need further confirm in the future as few wells have been drilled into the lower Permian strata. The source rocks are mainly composed of blackshale in Doushantuo Formation of Sinian and Qiongzhusi Formation of Lower Cambrian, which have been confirmed as high-quality source rocks, with total organic carbon content of 2.1%–3.65%, kerogen type of I, hydrocarbon generation intensity of $(18–67) \times 10^8$ m^3/km^2, and over-matured degree of organic matter in source rocks. The potential reservoirs mainly include bioclastic limestone in Baota Formation of Middle Ordovician, oolitic limestone in Middle-Lower Cambrian and algal dolomite in Dengying Formation of Upper Sinian (Guohui et al. 2000). The caprocks are mainly developed as local interlayered caprocks, including shale in Qiongzhusi Formation and mud shale in Yanwangbian Formation of Lower Cambrian, shale in Longmaxi Formation of Silurian and so on. The lower combination can be regarded as an important target for the next risk exploration in the northern Longmenshan area.

② The control of structural evolution on hydrocarbon accumulation

The deformation of northern Longmenshan is mainly controlled by the imbricated thrust nappe in Longmenshan, which causes the severe deformation in the inner Longmenshan orogenic belt, and also shortens the crust in front of Longmenshan mountain in Sichuan Basin, forming a series of westward thrust structures.

Qingchuan Fault and Liangshui-Linyansi Fault in the northern Longmenshan area had been formed in the Indosinian period, and directly connected with the surface. The oil and gas preservation in north Longmenshan area, especially in the back mountain belt, is seriously damaged by these two faults, with multi-period activities in the late-stage tectonic movements. Besides, the sealing capacity of the trap was generally damaged by the density exposed faults and seriously broken nappe rocks with strong stress in the nappe process due to the long nappe distance in the structural belt. The shallow structure in the northern Longmenshan area is an imbricate thrust system composed of fine clastic rocks in Silurian, Ordovician and Cambrian, with Majiaogba fault as the front fault. This imbricate thrust system may be a duplex structure developed with denudation of roof faults, which shows the strong tectonic action. And it can be seen that the structural deformation of the original system under the nappe system is strong with the duplex structure in hierarchy construction, in which the roof faults are converged in the detachment layers of Cambrian and the floor faults are converged in the basement of Pre-Sinian. It can be concluded that the preservation conditions of oil and gas have been seriously damaged with strong tectonic actions in the west area of Majiaoba-Zhuyuanba Fault: the pale oil reservoir in Sinian is seriously damaged by whole uplift and denudation in Ningqiang area, Shanxi province. The gravelly asphalt in Changjianggou Formation of Lower Cambrian, Qingyuan area of Guangyuan should come from the paleo oil reservoir in Sinian, while asphalt veins largely exposed in Kuangshanliang area are formed due to the faults and fractures developed in the late stage. The preservation conditions of oil and gas in the area became worse with reactivities and fracture of fault system, especially after the Wenchuan earthquake.

The area of the Eastern Majiaoba-Zhuyuanba Fault zone is less affected by the early tectonic movements due to its close to the basin. A series of traps are developed mainly by the tectonic movements in late stage. Moreover, the preservation conditions of oil and gas are not obviously affected by the undeveloped fractures which are all buried underground and generally disappeared in the detachment layers of Jialingjiang Formation and Leikoupo Formation of Middle-Lower Triassic. The preservation conditions of oil and gas in the west area of Majiaoba-Zhuyuanba Fault are relatively poor and not favorable for the hydrocarbon accumulation due to the strong tectonic actions and faults going directly to the surface. The unsealing of faults and the poor preservation conditions of oil and gas in Feixianguan Formation are performed by the repeated well leakage through actual drilling results. While the oil and gas preservation conditions in the eastern area of Majiaoba-Zhuyuanba Fault are relatively good with high-quality sealing in gypsum-salt layers of Jialingjiang Formation and Leikoupo Formation, which is caused by the weak tectonic actions and disappearance of fractures in the detachment layers of Jialingjiang Formation and Leikoupo Formation of Middle-Lower Triassic.

According to the data of Shuangtan 3 in the northern Longmenshan area, the source rocks of Permian began to subside continuously after sedimentation, and reached the mature stage to generate oil in the burial depth of about 4000 m in Early Triassic (Fig. 4.27). During the Middle Triassic to Late Jurassic, crude oil was cracked and a large amount of gas was generated, and the natural gas was migrated into reservoirs. However, the oil and gas reservoirs were developed by readjustment through the destruction of previous traps caused by the structural uplift in the early Cretaceous.

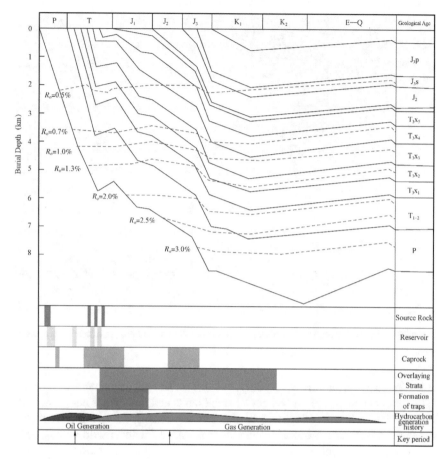

Fig. 4.27 Thermal evolution history of source rocks and hydrocarbon accumulation events of Shuangtan 3 in northern Longmenshan

At the end of the Late Triassic, the backward expansion superposition of three imbricate structures were included in structural deformation of northern Longmenshan area. After the deformation at the end of Late Triassic, the raised imbricate structures were denuded violently, and the Jurassic strata with great thickness were deposited in and around Sichuan Basin. During the Cenozoic deformation period, another sharp shortening deformation was developed in Longmenshan area with the rapid uplift of Tibetan Plateau under the impact of the collision between the Indosinian plate and the Eurasian plate. Two parts are mainly included in the deformation, which are the the imbricated superposition of two structural wedges in the deep strata and the corresponding broken deformation of the upper strata in Early Paleozoic (Fig. 4.28).

In the process of nappe, the small faults and fractures near the large faults are also significance to the migration and preservation of oil and gas. In this condition, the

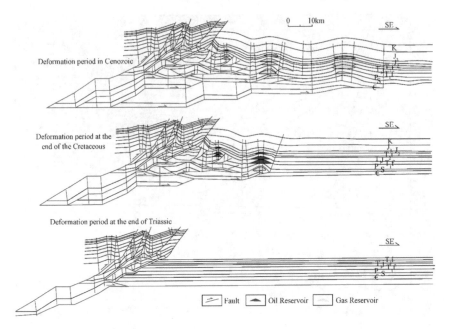

Fig. 4.28 Structural evolution and hydrocarbon accumulation model of northern Longmenshan area

migration in the reservoirs is very important for oil and gas accumulation, although the migration distance of oil and gas by small faults and fractures is not farther than by large faults. The development of dissolution pinpores, pore-cave and fracture systems, as well as the characteristics of multi-stage fractures with cutting pores can be seen in the Well Kuang 2 coring intervals of saccharoidal dolomite of Qixia Formation of Permian and field investigation. Such migration mode of oil and gas can be proved by the filling of black asphalt in some dissolution pinpores, a few pore-cave and micro fractures.

4.3 Accumulation Model of Deep Oil and Gas

Different accumulation models and distribution controlling factors are developed in deep reservoirs with different lithologies. The deep carbonate reservoirs can be divided into primary and secondary oil and gas reservoirs according to the formation mechanism. Primary reservoirs can be divided into single-source and single-stage reservoir, as well as multi-source and multi-stage reservoir. Secondary reservoirs can be divided into three types according to the geological evolution. The large-area accumulation model of deep clastic oil and gas depends on the scale and sufficiency of hydrocarbon source kitchen, reservoir continuity, trap integrity and effectiveness of cap rock. While the deep volcanic reservoirs are near source with the distribution in or near the hydrocarbon generation depression and formed with the conditions of "first come, first served".

4.3.1 Hydrocarbon Accumulation Model of Deep Carbonate Rock

The characteristics of multi-stage hydrocarbon generation and accumulation are developed in marine basins onland in China. The carbonate rock accumulation models and reservoir types are also diverse due to the multi-stage accumulation, including self generation and self preservation, lower generation and upper preservation, as well as complex intersection accumulation. According to the types of trap, the reservoirs can be divided into structural reservoir, lithologic reservoir, stratigraphic reservoir and complex reservoir. According to the classification of history-genesis in the oil and gas system, the reservoirs can be divided into primary oil and gas system, residual oil and gas system, secondary oil and gas system and destructive oil and gas system et al. The processes of multi-stage accumulation, adjustment, transformation and reenrichment of oil and gas are widely developed in marine carbonate reservoirs with multi-stage hydrocarbon generation in source rocks (Fig. 4.29). Therefore, the reservoirs can be divided into primary and secondary oil and gas reservoirs according to the formation mechanism.

The primary reservoir is mainly accumulated by the direct migration of oil and gas generated from the source rocks, with the relatively stable or little changed trap position and formation after the reservoir formation. The secondary reservoir is developed by the transformation, readjustment and remigration of oil and gas in the primary reservoir. Among them, some carbonate reservoirs in Tabei and Tazhong areas of Tarim Basin, natural gas reservoirs in Paleozoic of Ordos Basin, and natural gas fields containing hydrogen sulfide in Changxing-Feixianguan Formation of East and North Sichuan Basin are all primary reservoirs (Shuichang et al. 2006; Zhao Wenzhi et al. 2006a, b); while the gas fields group of Carboniferous in East Sichuan, Weiyuan Gasfield of South Sichuan, Hetianhe Gasfield of Tarim Basin are secondary gas reservoirs with natural gas generated from crude oil cracking, which the paleo oil reservoirs are deeply buried and cracked, and then redistributed and accumulated in different places.

(1)   Primary Oil and Gas Reservoirs

The primary oil and gas reservoirs can be divided into reservoirs of single-source and single-stage and reservoirs of multi-source and multi-stage. At least three times of large-scale hydrocarbon filling are developed in both Lungudong area and condensate gas reservoir in Tazhong I belt (Fig. 4.30) in Tarim Basin, with oil and gas come from source rocks in Cambrian and Ordovician respectively. The filling time are late

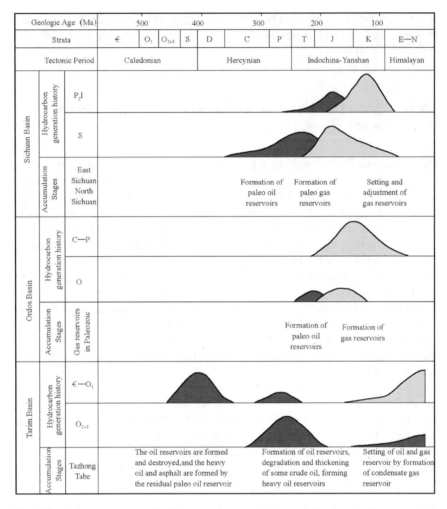

Fig. 4.29 Hydrocarbon generation history and accumulation periods of major Source rocks in marine basins in China

Caledonian, late Hercynian and late Himalayan, which the most important effective accumulation periods are developed in late Hercynian and late Himalayan. In late Himalayan, the reservoirs are mainly filled with the gas, which has a certain degree of gas cut and transformation on the oil reservoirs in late Hercynian. However, the carbonate reservoirs in late Hercynian are well preserved as most of the reservoirs in the central and western areas of northern Tarim are not affected by the natural gas filling in Himalayan period. For example, the Yingmai 2 reservoir in Yingmaili area is an ancient reservoir formed 250 million years ago, which is filled and formed in Late Hercynian, and effectively preserved for the stability of later structures. In the Late Caledonian, the oil and gas are mainly distributed in the middle and eastern part

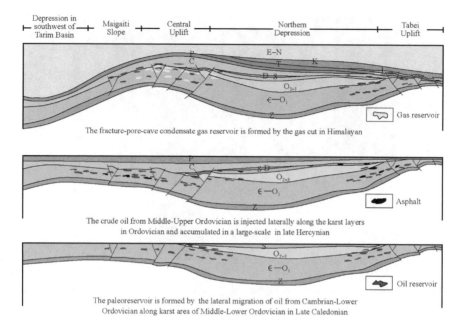

Fig. 4.30 Multi-stage accumulation model in the platform area of Tarim Basin

of the Middle-North Tarim Basin. However, the reservoirs are basically destroyed by the uplift of Tabei area in early Hercynian and only partly preserved in the central Tarim Basin.

(2) Secondary Oil and Gas Reservoirs

Secondary oil and gas reservoirs are widely distributed in marine basins in China. The preservation conditions of primary oil and gas reservoirs are changed with the constantly changes of form and location of traps, through the superposition of different types basins and the compound action in different geological processes during the evolution process in marine basins in China. So when the accumulated oil and gas in the reservoirs are in the adjustment process of constant redistribution, the complexities in phase state and distribution of oil and gas are formed by crude oil cracking through destruction and deep burial. According to the current exploration practices, secondary oil and gas reservoirs can be divided into three models (Fig. 4.31): (1) Several small reservoirs are decomposed from large-scale paleo-reservoirs in the late stage. Large-scale accumulation of oil and gas near Xiang 3 was developed in Tabei area in late Hercynian. In the late stage, several oil and gas reservoirs were formed by the redistribution of oil and gas by the anti-dip of strata. For example, Hudson Oilfield is a secondary reservoir formed by later structural inversion. (2) Several small reservoirs are decomposed from ancient reservoirs in situ. Reservoir in Carboniferous of East Sichuan is a typical representative. It was a paleo oil reservoir in late Early Jurassic. Owing to the rapid deposition of overlying strata,

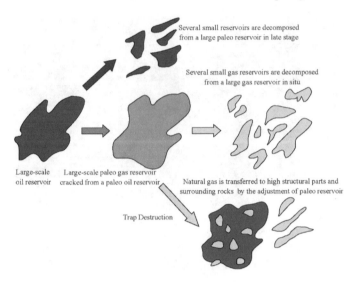

Fig. 4.31 Three models of secondary adjustment and accumulation of marine carbonate reservoirs in China

a large-scale paleo gas reservoir was formed in situ by complete decomposition of the paleo oil reservoir at the end of Cretaceous. And then a serious of small gas reservoirs were developed by the adjustment and destruction of the paleo gas reservoir due to the tectonic movements in Himalayan. (3) Oil and gas are reaccumulated in different places after the oil cracking and gas generation of the paleo reservoir. As a typical representative, large amount of liquid hydrocarbons in Weiyuan Gasfield were formed from Cambrian source rocks in ancient depression in Caledonian. At the same time, Ziyang uplift was formed, which was developed as an important trap for oil and gas accumulation. The gas reservoir was formed by the cracking of ancient reservoir in Yanshanian. Finally, Weiyuan Gasfield was formed in the high structural part, with the adjustment in Himalayan.

4.3.2 Large-Area Hydrocarbon Accumulation Model of Deep Clastic Rock

(1)   Scale and sufficiency of hydrocarbon source kitchen

The accumulation of oil and gas with large area in deep clastic reservoir is developed with the superposition of hydrocarbon source kitchens and the thrust belts over the hydrocarbon generating center and late generation and expulsion periods in hydrocarbon. Kuqa Depression in Tarim Basin is a foreland depression with terrigenous clastic deposits of Mesozoic and Cenozoic with the maximum buried depth of more than 8000 m. The strata of Quaternary, Neogene, Paleogene, Cretaceous, Jurassic and Triassic are developed from top to bottom, among which the strata of Jurassic and Triassic are the main source layers. Five formations are mainly developed in source rocks of Jurassic and Triassic, which are the Huangshanjie Formation (T_3h), the Taliqike Formation (T_3t), the Yangxia Formation (J_1y), the Kizilenur Formation (J_2k) and the Qiakemake Formation (J_2q) from the bottom to the top. Among them, the strata in Triassic are mainly lacustrine mudstone with carbonaceous mudstone at the top; the strata in Jurassic are mainly swamp-lacustrine coal-bearing depositions, and the coal layers are mainly developed in Yangxia Formation and Kizilenur Formation, with a thickness of 6–29 m and a maximum thickness of 66 m; the coal-measure source rocks in Jurassic are widely distributed in Kuqa Depression, with large thickness and high organic matter abundance. The main gas source rocks in Kuqa Depression are organic matters of III-type. Large amount of natural gas is produced in the process of hydrocarbon generation in the thermal evolution, especially in the middle-high evolution stage. The source rocks of Triassic are mainly composed of lacustrine mudstone with high organic matters abundance, and the organic matters are mainly III-type and II-type, which are similar to the source rocks in Jurassic coal-measure source rocks, producing gas. Especially after entering the middle-high evolution stage, the gas hydrocarbon generation rate increases obviously. The source rocks in Triassic and Jurassic are developed with superposition. The total thickness of the two sets of source rocks is 400–1700 m, the area is 2.16 $\times 10^4$ km^2, the gas generation intensity is $(350–400) \times 10^8$ m^3/km^2, and the gas production is 204×10^{12} m^3. The hydrocarbon source kitchens are superimposed (Fig. 4.32) and supply hydrocarbon; the thrust belts are superimposed on the hydrocarbon generation center, and the thickness of the source kicthens increases by 3–5 times. In addition, the hydrocarbon generation and expulsion of source rocks are mainly in the Late Cenozoic, with fracture communication, as well as continuous and efficient filling of nature gas.

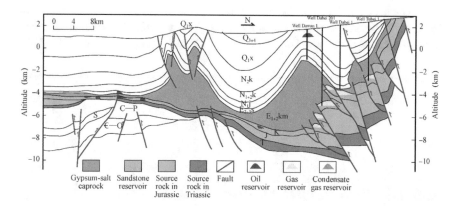

Fig. 4.32 The superposition of source kitchens and natural gas accumulation models in Kuqa Depression

In Kuqa Depression, the natural gas generated in gas source rocks of Triassic and Jurassic can only be accumulated in the Cretaceous reservoirs by the migration of fractures, as the huge-thick mudstones of Shushanhe Formation of Cretaceous, Xilazha Formation and Qigu Formation of Jurassic are developed between the Cretaceous reservoirs and Triassic and Jurassic gas source rocks of. Four types of nature gas accumulation models migrated by faults are summarized based on the research on the gas accumulation conditions of typical structures in Kuqa Depression. Such as the gas migration and accumulation models composed of under-salt faults and crossing-salt faults (not connect with traps), only composed of under-salt faults, composed of under-salt faults and faults crossing the top of traps, and only composed of crossing-salt faults.

(2)   Reservoir continuity

The reservoir can be maintained by rapid burial in late stage with the large-area development of delta sandstone. Taking the Cretaceous Bashijiqike Formation in Kelasu thrust belt of Kuqa Depression as an example, the paleogeomorphology is relatively flat with relatively weak tectonic activities in the sedimentary period of Bashijiqike Formation Member 2. The provenance mainly comes from Northern Tianshan and extends to the basin. The large-scale lobate sandbodies are developed in the research area due to the abundant sediment supply and strong hydrodynamic environment which cause the extension of delta from northern piedmont to basin. The braided-river delta plain composed of a serious of braided channels rather than single braided channel is formed by multiple material source mouths of north part (Fig. 4.33). The total thickness of the sediment can reach more than 300 m, and the sand bodies is relatively thick. The sand bodies of delta front are vertically superposed and horizontally connected, with reservoir thickness of 200–400 m and area of about 1.8×10^4 km^2. In addition, the long-period shallow burial and the rapid deep burial in late stage are conducive to the maintenance of the primary pores; meanwhile, the fractures are well developed by tectonic activity. As a result, the effective reservoir is still developed at the burial depth of 6000–8000 m, with the porosity of 5%–10%.

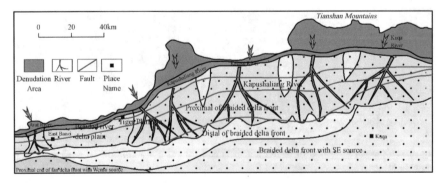

Fig. 4.33 Sedimentary facies map of Bashijiqike Formation Member 2 in Cretaceous

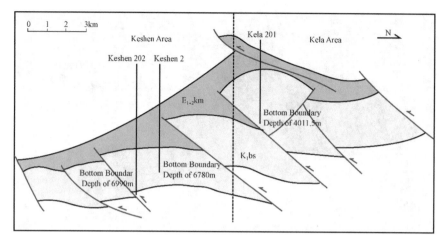

Fig. 4.34 Structural characteristics of Keshen Section in Kelasu structural belt of Kuqa Depression

(3) Trap integrity

The traps are completely developed with the top and side sealing of gypsum-salt layers, and the development of large-scale under-salt thrust structure. During the strong orogeny of South Tianshan in the middle-late Kuqa period, the traps with the strong oil and gas filling are the anticlines distributed in rows and belts by a serious of basement involvements and cover-decollement thrust faults in under-salt layers, which are formed in the large-scale thrust nappe structures. The thrust imbricate structures are developed as 5–10 rows, with the scale distribution in space and traps. The effective traps such as Kela 2 anticline and Keshen faulted anticline are complete structural traps with the sealing of gypsum-salt layers in top and side (Fig. 4.34).

(4) Caprack Effectiveness

High-quality caprocks of thick gypsum-salt rocks are distributed in large area and sealed with high strength. The differences in migration and accumulation of oil and gas are controlled by the fractures and sealing of gypsum-salt rocks in Kuqa Foreland Basin, as well as the orderly distribution of faults-caprocks combination in time and space, which are all caused by the brittle-plastic evolution of gypsum-salt rocks and the development of thrust faults. Most of the traps in northern Kela area of Kelasu rich-gas structure belt are failed, as the traps in the Kelasu fault northern zone are developed with the broken fault-caprock combination (Fig. 4.35), which are formed by the shallow burial of Paleogene gypsum-salt rocks, brittle deformation and long-term activities of thrust faults. In the southern Keshen area, the reservoirs are developed with plasticity and deep burial depth in the large-scale accumulation period of natural gas. As a result, the conditions of large-scale gas fields are developed by the formation of unbroken fault-caprock combination as it is difficult for the faults in the late stage to cause substantial damage on the traps. While the traps between the

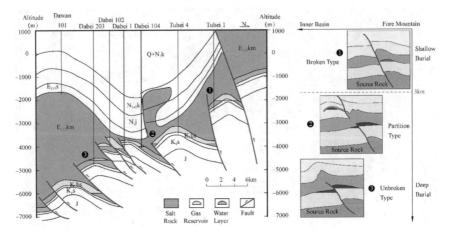

Fig. 4.35 Sealing patterns of caprock

upper and lower walls close to the Kelasu fault zone are between them. The partition fault-caprock combination is developed by the early-stage fracture and late-stage sealing of gypsum-salt layers. A small amount of crude oil in the early stage and a large amount of natural gas in the late stage are captured and accumulated by the gypsum-salt caprocks in Dabei area, with the burial depth of far more than broken 3000 m, and the formation time of caprock sealing in early-middle Kuqa period. The fault-caprock combinations of unbroken, partition and broken are orderly developed in Kelasu structure belt from the south to the north. The thickness of Paleogene gypsum-salt layers in the west of thrust belt is 500–3500 m; and the thickness of Neogene gypsum-salt layers in the east is 500–2000 m, with a total area of 1.9 × 10^4 km^2. The breakthrough pressure of gypsum-salt layers is as high as 15–20 MPa. The better property of sealing is developed in gypsum-salt layers below 3000 m with complete plasticity and unbroken by faults.

4.3.3 Hydrocarbon Accumulation Model of Deep Volcanic Rocks Near Source Rocks

The primary condition for the formation of volcanic reservoirs is the association with source rocks as the volcanic rocks can not generate organic hydrocarbons. That is, the good matching relationship between volcanic reservoirs and source rocks in sedimentary layers can only be formed by the location of volcanic rocks developed in the source rocks, above or below the source rocks, or the hydrocarbon generating depressions nearby. Sufficient oil source supply is the necessary condition for the formation of volcanic reservoirs. The favorable volcanic-sedimentary sequence accumulation is developed by the interbedding of volcanic layers and sedimentary layers in main continental petroliferous basins in China, such as Songliao Basin,

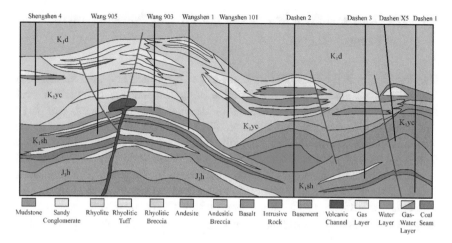

Fig. 4.36 Gas reservoir profile of Anda Area in Xujiaweizi, Songliao Basin

Bohai Bay Basin and Junggar Basin. Therefore, the deep volcanic accumulation has the characteristics of near source rocks, with two types of oil and gas accumulation, such as lithologic and stratigraphic reservoirs, both of which can be distributed in a large scale.

The conditions of source, reservoir, cap, migration, trap and preservation, as well as favorable allocation in time and space must be developed in volcanic oil and gas reservoirs, but the accumulation law and distribution are specific. The types of volcanic reservoirs are various and mainly developed as structural-lithostratigraphic reservoirs. Such as the lithologic gas reservoirs of volcanic rocks in Xujiaweizi depression of Songliao Basin are developed with superposition of many gas reservoirs (Fig. 4.36), with no unified gas–water interface, poor connectivity, and gas column exceeding the structural amplitude; the stratigraphic reservoirs in Permian of northwestern margin of Junggar Basin are formed by the development of pore-cave-fracture reservoirs in volcanic weathering crust in all kinds of rocks, and the oil and gas accumulation are controlled by unconformity. In the near-source combination, the oil and gas generated from source rocks can be contacted with reservoirs in maximal degree as the source rocks are located in the upper and lower or lateral margin of the volcanic reservoirs, and the volcanic reservoirs are distributed in or near the hydrocarbon generation depression. Generally speaking, the condition of "first come, first served" in the volcanic rocks is developed in the near-source combination, which is most conducive to the accumulation of oil and gas.

4.4 Distribution Law of Large Oil and Gas Fields in Deep Carbonate Rocks

The distribution and enrichment of multi-stage adjustment and transformation of deep oil and gas are related to the selection of exploration direction and the evaluation of favorable exploration zones. Multi-stage tectonic movements can play a role of destruction and transformation on oil and gas accumulation, as a result, the accumulation and distribution of deep oil and gas are complex. The formation of deep-large oil and gas fields is controlled by the large-scale source kitchens, and the accumulation of oil and gas is largely controlled by the paleogeographic background of structure and lithofacies. Although the accumulation of deep oil and gas is complex, it is still controlled by source kitchen. The deep strata in superimposed basin are developed with many "exploration golden zones", and the prospect of deep exploration is optimistic, as the distribution of large oil and gas fields is controlled by main source kitchens, and the accumulation of oil and gas is controlled by slope, paleo-platform margin and paleo-fault zone.

4.4.1 Source Control of Deep Oil and Gas Distribution

The so-called source-control refers to that the main oil and gas reservoirs formed in each oil and gas accumulation layer of superimposed basin are distributed within the effective source kitchen or the area closely related to the source kitchen. Two types of hydrocarbon source kitchens in the superimposed petroliferous basins are mainly developed in China: one is source kitchen of kerogen, which is the source kitchen formed by the thermal cracking of organic matters (kerogen) in the source rocks; the other one is source kitchen with liquid hydrocarbon cracking, which is formed by the cracking of dispersed liquid hydrocarbon in the paleo oil reservoir or remained in the source rocks during the stage of high-over mature, and mainly generates gas. Exploration practice has proved that both types of source kitchen can supply hydrocarbons on a large scale and are highly efficient source kitchen. At present, the oil and gas discovered in strata of superimposed petroliferous basin are mainly controlled by these two kinds of hydrocarbon source kitchens.

Three hydrocarbon generating centers (Fig. 4.37) are formed in the source rocks in Cambrian-Ordovician in Tarim Basin. Manjiaer Depression is the main hydrocarbon generating center, with a distribution area of about $(10–12) \times 10^4$ km^2, and a maximum hydrocarbon generating intensity of 160×10^8 t/km^2; the next is Awati hydrocarbon generating center, with an area of about $(8–10) \times 10^4$ km^2, and a maximum hydrocarbon generating intensity of 40×10^8 t/km^2. The Tabei Uplift lies in the north margin of hydrocarbon generating centers in Manjaer and Awati, with high enrichment of oil and gas. Many large-scale oil and gas fields, such as Lunnan-Tahe, Halahatang and Yingmaili, have been explored, with oil and gas reserves of more than 20×10^8 t oil equivalent; the Tazhong Uplift lies between

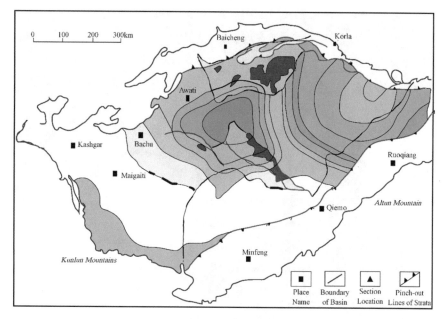

Fig. 4.37 Superimposition of hydrocarbon fields and source rocks of Ordovician in Tarim Basin

these two hydrocarbon generating centers, with a discovered oil and gas reserves of nearly 10×10^8 t oil equivalent. The distribution of oil and gas in Sinian-Cambrian of the Sichuan Basin is obviously controlled by hydrocarbon source kitchens even in the ancient strata (Fig. 4.38). In recent exploration within the control range of the source rock thickness center, the large gas field of Sinian-Cabrian in Gaoshiti-Moxi area is newly discovered as the oldest producing layer and the largest single reserves in China.

The above two types of source kitchens are interdependent. The source rocks kitchen of kerogen are the basis, and the source rocks of liquid hydrocarbon cracking are the derivatives of kerogen-type source rocks. The hydrocarbon generation time is early in kerogen-type hydrocarbon source kitchen, with not only oil generation but also gas generation; while the hydrocarbon generation time is late in hydrocarbon source kitchen of liquid hydrocarbon cracking, mainly with gas generation. For the large oil and gas fields, especially large carbonate gas fields, it is generally supplied by both types of hydrocarbon source kitchen. Such as Manjiaer sag in Tarim Basin, which is not only the center of kerogen-type source kitchen but also the center of source kitchen with residual and dispersed liquid hydrocarbon cracking. The control and contribution of this kind of hydrocarbon source kitchen is confirmed by the major breakthrough of Well Gucheng 6 in Gucheng area.

The formation of large oil and gas fields is controlled by three types of near-source accumulation combinations. Three types of near-source accumulation combinations are developed in marine carbonate oil and gas fields in China, such as the rift-platform margin accumulation combination, "burger" depression accumulation combination

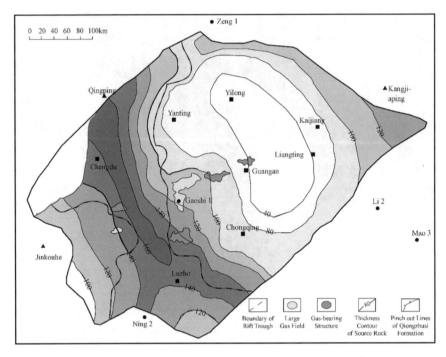

Fig. 4.38 Superposition of source rocks and large gasfield in Gaoshiti-Moxi area of Sichuan Basin

and depression-paleouplift accumulation combination. The source-reservoir-caprock conditions in different accumulation combinations are different by the control of paleostructure-sedimentary background variance (Table 4.3).

(1)   Rift- platform margin accumulation combination

This kind of accumulation combination is mainly developed near the platform margin belt on both sides of the rift trough in craton basin, which is characterized by the distribution of the main gas reservoirs along the rift trough. The gas sources mainly come from the high-quality source rocks in the rift trough under the reservoir, and the high-quality source rocks in or above strata. The large-scale reservoirs are developed with excellent physical properties, which are mostly reef beach or mound beach. The faults, unconformities and fractures are the main passage systems with the direct contact or lateral contact of source rocks and reservoirs. The composite traps of structure and lithology-structure are developed with relatively large closed area and height, which are conducive to large-scale accumulation of oil and gas. The accumulation model can be called transition model, with the accumulation characteristics of near-source self-generation and self-storage, upper-generation and lower-storage or lower-generation and upper-storage such as the reef beach of Changxing Formation in Sichuan Basin, mound beach of Dengying Formation in Sichuan Basin and reef

Table 4.3 Three types of near-source accumulation combination and key elements of marine carbonate gas fields in China

Accumulation combination	Source rock	Reservoir	Passage system	Model	Example
Rift-platform margin combination	High-quality source rocks in rift area	Reef (mound) beach bodies	Unconformity and fault		Reef-beach in Changxing Formation of Sichuan Basin, mound-beach in Dengying Formation of Sichuan Basin and Lianglitage Formation of Tarim Basin
"Burger" combination in depression	Wide covered source rocks	Stratiform reservoirs in weathering crust and grain beach inside platform	Unconformity and porous reservoir		Ordovician in the Ordos Basin; Longwangmiao Formation, grain beach inside plarform of Changxing Formation and Leikoupo Formation in the Sichuan Basin; weathering crust of Yingshan Formation in the Tarim Basin
Depression—paleouplift Combination	Near the hydrocarbon generating center	Stratiform paleokarst	The network transportation of fracture and unconformity		Slope of Tabei area and North Slope of Tazhong are in the Tarim Basin

beach of Lianglitage Formation in Tarim Basin, which all have this kind of accumulation combination in rift platform margin. For the mound-beach gas reservoir of Dengying Formation in Sichuan Basin, the source rocks mainly come from the high-quality source rocks in Deyang-Anyue rift trough and overlying Qiongzhusi Formation as well as the source rocks in Member 3 of Dengying Formaion, and the reservoirs are developed as mound-beach in Member 2 and 4 of Dengying Formation in platform margin belt. The source rocks are directly or laterally connected with the reservoir, and the favourable passage systems are unconformity, faults and fractures. The traps of structure and lithology-structure are developed, with the characteristics of near-source self-generation and self-storage, upper-generation and lower-storage.

(2)   "Burger" in depression accumulation combination

This kind of accumulation combination is mainly developed near the regional unconformity, whose characteristics are that the gas sources mainly come from the source rocks above the unconformities, the reservoirs mainly develop in the large-area weathering crust under the unconformities, the surface or oblique contact is developed between the source rocks and the reservoirs, the main passage system is the unconformity, and the oil and gas migration power mainly comes from the source-reservoir pressure difference. The higher fluid pressure is formed by the large amount of hydrocarbon generated from organic matters in the overlying source rocks. The oil and gas enter the reservoir and accumulate by overcoming the buoyancy, when the fluid pressure in overlying source rocks is larger than that in the underlying reservoirs. The accumulation model can be called as backward charging, with the characteristics of near-source upper-generation and lower-storage, such as the Ordovician gas reservoir in Ordos Basin, the weathering crust gas reservoir of Yingshan Formation in Tarim Basin and Leikoupo Formation in Sichuan Basin. The excellent source rock-caprock combination is developed in Ordovician gas reservoirs of Ordos Basin, whose source rocks and caprocks is the strata of overlying Mesozoic coal-measure mudstone. The reservoir is developed as the karst reservoirs of weathering crust on the top of Majiagou Formation of Ordovician, with direct or oblique contact between the source rocks and reservoirs. And the oil and gas accumulation is developed with the characteristics of near-source upper-generation and lower-generation, with the favourable passage system composed of unconformity and dissolved fracture-cave. The large gas field is developd by the large-scale oil and gas accumulation in Ordovician of Ordors Basin with high-quality source rock-reservoir-caprock combination.

(3)   Depression-paleouplift accumulation combination

This kind of accumulation combination is mainly developed in the upper part and surrounding slope belt of paleouplift, which is characterized by the the gas source mainly comes from the source rocks in the lower part or flank of paleouplift, the reservoir mainly are the large-scale layers with dissolution fractures and caves in the upper and surrounding of paleouplift, the source rocks and reservoirs are in oblique contact with each other in the same layer or connected by gas source fractures, the major passage systems develop as the fracture-cave network in the reservoir and the fracture system connecting the gas source, and the migration power of oil and gas mainly comes from buoyancy storage. The large-scale hydrocarbon generated by source rocks in the low part of the uplift is accumulated and transferred in the "big cave" by the passage system of fracture-cave network, and buoyancy storage is developed at the same time. The hydrocarbon will migrate from the low part to the higher part along the fracture-cave network on the slope when the buoyancy reaches a certain degree and can overcome the upward migration resistance, and this process can be called "buoyancy storage, lift transfer". Repeated so many times, the large gas field can be formed by the large-scale oil and gas accumulation in fracture-cave reservoirs in the upper part and surrounding slope of paleouplift. For example, a large-scale layered fracture-cave network is developed from the higher part of the uplift to the lower part of the slope in Ordovician of Tabei, in which large-scale layered oil and gas reservoirs are developed. The gas reservoir is mainly developed as fracture-cave reservoir, and the source rocks are the mudstone in Cambrian and middle-lower Ordovician, with excellent source rock-reservoir combination. The near-source accumulation is developed with the characteristics of "lift", and the favourable passage system composed of dissolved fractures and caves.

4.4.2 Hydrocarbon Accumulation Controlled by Paleouplift, Slope, Paleoplatform Margin and Paleofracture

As a general rule, the most important places for oil and gas enrichment are the large-scale paleouplift and paleoslope developed for a long stage. In addition to the formation of large-scale structural, lithologic and stratigraphic traps, the large-scale oil and gas accumulation can also be developed by the control of paleouplift and paleoslope, such as the absorption of oil and gas, and the control on the development and distribution of large-scale reservoin. Such as the large-scale carbonate oil and gas fields found in the Tabei, Tazhong uplift and slope areas of Tarim Basin, the Sinian-Cambrian gas fields found in the central Sichuan paleouplift of Sichuan Basin, the gas fields in Xujiahe Formation of Upper Triassic in central Sichuan, and the Upper Paleozoic Sulige Gasfield found in Ordos Basin, all of which are controlled by palaeouplift or palaeoslope. The margin belt of paleoplatform is a favorable part for the development of carbonate reef-beach reservoir. The multi-layer large-scale oil and gas accumulation can be formed by the platform-margin faults connection of the superimposed reservoirs of reef beach and clastic sandstone, with the superimposed deposition of large-scale river–lake delta on the platform-margin belt. Taking the platform margin belt of Longgang in Sichuan Basin as an example, many gas bearing strata have been found in reef breach of Permian and Triassic, carbonate weathering crust in Leikoupo Formation of Triassic, and Xujiahe Formation in Triassic. Fault is not only an important passage for oil and gas migration but also a favorable position of hydrocarbon migration and accumulation. On the one hand, the long-developed paleofaults can form fracture zones, and on the other hand, they are beneficial to deep hydrothermal activities, so that the physical properties of the reservoir can be improved. For example, the carbonate reservoirs found in Tazhong area of Tarim Basin and Halahatang area on the south margin of the North Tarim Basin are all related to fault activities.

(1) The distribution of oil and gas reservoirs in three major basins are controlled by the paleouplift and slope

Oil and gas exploration with "paleouplift and paleoslope" is an important recognition in the long-term exploration practice of carbonate reservoirs. The typical oil and gas fields are developed as Tazhong (Fig. 4.39), Tahe, Halahatang (Fig. 4.40), Yingmaili, Hetianhe et al. The main production layers are Yijianfang Formation, Yingshan Formation and Lianglitage Formation, with a proved reserve of 37.08×10^8 t, a third-grade reserve of 51.76×10^8 t, and main oil source is in Cambrian.

In the early exploration of Sichuan Basin, Ziyang-Weiyuan Gasfield was discovered in the central Sichuan Basin. During the 12th five-year plan period, two paleouplifts, which are Caledonian synsedimentary paleouplift of Longwangmiao Formation and Luzhou-Tongjiang paleouplift of Dongwu period, are newly discovered, depending on the progress of seismic technology and comprehensive geological knowledge. The distribution scale of grain beach reservoirs in Longwangmiao Formation is controlled by the former paleouplift, and the distribution scale of karst reservoirs in Maokou Formation of Middle Permian is controlled by the latter paleouplift.

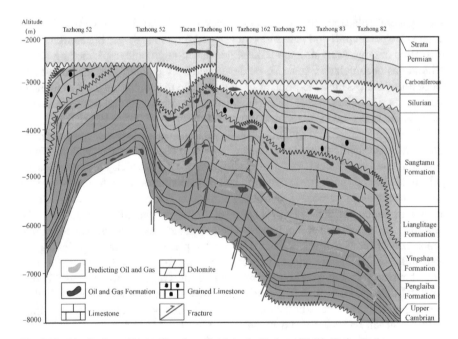

Fig. 4.39 Distribution of large oil and gas fields in the Tazhong Uplift, Tarim Basin

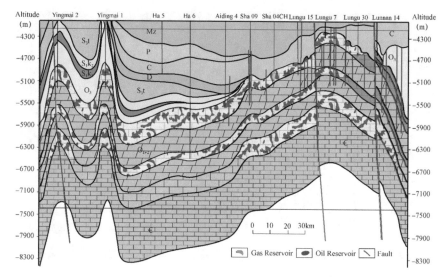

Fig. 4.40 Oil and gas accumulation profile of paleouplift in Tabei area, Tarim Basin

The discovery of paleouplifts plays a key role in the discovery of Anyue Gasfield and the breakthrough of Middle Permian exploration.

(2)   The distribution of large-scale reef beach oil and gas fields controlled by the paleoplatform margin

The paleoplatform margin belt is a favorable part for the carbonate reef-beach reservoir. The multi-layer largescale oil and gas reservoirs can be formed by the platform-margin faults connection of the superimposed reservoirs of reef beach and clastic sandstone, with the superimposed deposition of large-scale river–lake delta on the platform-margin belt. Taking the platform margin belt of Longgang in Sichuan Basin as an example, many gas-bearing strata have been found in reef breach of Permian and Triassic, carbonate weathering crust in Leikoupo Formation of Triassic, and Xujiahe Formation in Triassic. Taking reef beach gas reservoirs of Kaijiang-Liangping and Pengxi-Wusheng in trough -platform margin belt as examples (Fig. 4.41), under the control of high-energy environment, five reef beach bodies of 3×10^4 km^2 in platform margin belt are developed, with 68 reef bodies with an area of 5500 km^2. The oil and gas accumulation is characterized by "one reef, one beach and one reservoir", and distributed in the shape of "bead" along the margin of the platform, with a reserve abundance of $(4–40) \times 10^8$ t/km^2.

(3)   Oil and Gas Enrichment Controlled by Paleofaults

The carbonate reservoirs found in Tazhong area and Halahatang area on the south margin of Tabei the Tarim Basin are all related to fault activities (Shi shuyuan et al. 2015). Taking the paleokarst reservoir of Ordovician Yingshan Formation in Yueman

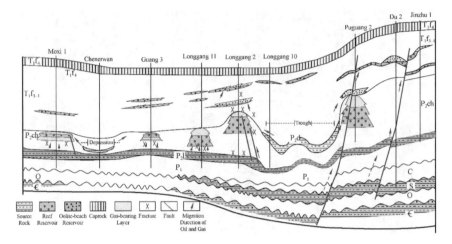

Fig. 4.41 Reef beach gasfields in the trough-platform margin belt of Kaijiang-Liangping and Pengxi-Wusheng

Block of Tabei as an example (Fig. 4.42), four hydrocarbon enrichment zones of fracture-cave (Yueman 1, Yueman 2, Yueman 3 and Yueman 4) are developed along the faults, with an oil-bearing area of 243.8 km^2, a geological oil reserve of 2741.4 × 10^4 t, a technical recoverable reserve of 575.20 × 10^4 t, an average daily oil production of 35 t for a single well, and a cumulative oil production of 15.74 × 10^4 t.

In a word, two types of hydrocarbon kitchens, such as hydrocarbon kitchen of kerogen and hydrocarbon kitchen of liquid hydrocarbon cracking are developed in deep marine carbonate strata, with large-scale hydrocarbon supply; and many "golden zones" for exploration are found under the control of paleouplift, paleoslope, paleoplatform margin and multiple successive faults. As a result, the exploration prospect of carbonate strata is optimistic with the large scale of oil and gas distribution and reserves, although the degree of oil and gas enrichment in different structures is different. In recent years, a prospect for deep oil and gas exploration is shown by the continuous breakthroughs of carbonate exploration in Tarim, Sichuan and the other basins. In particular, the newly discovered Cambrian gas field in Longwangmiao Formation of the central Sichuan Basin, is developed not only with a large-scale reserves which single reserve is up to 4404 × 10^8 m^3 but also with a high output in single well and good effect on production test. Ten wells with a daily production of more than one million cubic meters are drilled, with a maximum open flow capacity of 1035 × 10^4 m^3 and a daily production capacity of 480 × 10^4 m^3. The prospect of oil and gas is worth looking forward to by the great contribution to the development of China's oil and gas industry in the field of carbonate reservoirs with the deepening of geological knowledge and the progress of engineering technology.

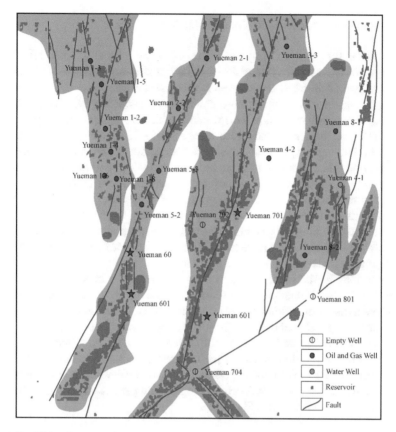

Fig. 4.42 Oil-bearing area of proved reserves in Ordovician of Yueman Block

4.4.3 Multi-exploration Golden Zones in Deep Strata

In recent years, the explorations in Sichuan, Tarim and Ordos basins show that
the carbonate reservoirs in China are developed with the characteristics of multi-
layer accumulation, especially in the deep strata. At present, the exploration depth
has exceeded 6000 m, far beyond the lower limit of exploration "golden zone"
proposed by Norwegian scholars. The theory of multi-exploration "golden zones" is
put forward by summarizing the petroleum geological characteristics of oil and gas
accumulation in multi-source, multi-reservoir and multi-stage, based on the multi-
cycle development of superimposed basins in China. Focusing on the characteristics of many sets of source rocks and reservoirs and multi-stage accumulation
in superimposed basins of China, many exploration "golden zones"can be developed by the multi-layer (zones) of oil and gas enrichment with the spatiotemporal
coupling of reservoir-forming factors. The optimistic exploration prospect and deep
exploration potential of superimposed basins in China are revealed by the theory of
multi-exploration "golden zones",

(1)   The Connotation of Multi-Exploration "Golden Zones"

In the 1970s, the concepts of hydrocarbon generation threshold, liquid oil window and dry gas stage, as well as the spatial distribution and temperature range of kerogen evolution were defined by the kerogen hydrocarbon generation model proposed by Tissot. Among them, the exploration "golden zone" defined by Norwegian scholars is mainly developed in the liquid oil window, with the underground temperature of 60–120 °C, and the corresponding Ro value of 0.6%–1.2%, which is the main depth range of oil and gas exploration. The marine strata developed in the deep layers of the superimposed basins in China are old, with high degree of thermal evolution in source rocks and sufficient evolution of organic matters. The hydrocarbon generation process is developed with the characteristics of "two peaks", as a large-scale oil generation is developed in the early stage, and a large scale the dispersed liquid hydrocarbons cracking gas generation by remaining in the source rocks after entering the later stage of high-over mature. At the same time, multi-stage and multi-layer source rocks are developed in the superimposed basins, and many source kitchens can be formed by the source rocks in the same layer. The history of hydrocarbon generation is also different in these source kitchens due to the different evolutions. The characteristics of multiple layers in the vertical, multiple belts and multiple areas in the plane are developed by the multi-phase accumulation of oil and gas, which are caused by the multi-layer development of source rocks, multi-stage and multi-source in hydrocarbon generation, and multi-period development of reservoirs in geological history period. Many exploration "golden zones" will be developed in superimposed petroliferous basins, if each layer controlling oil and gas enrichment is regarded as an exploration "golden zone" (Fig. 4.43).

The strata in Sinian to Middle Triassic in Sichuan Basin were marine carbonate sedimentary sequence. After the Late Triassic, the basin was closed due to the uplift of surrounding mountain system, and strata were gradually turned into continental clastic sedimentary sequence. At least five exploration "golden zones" in the basin are developed from the bottom to the top, such as the strata of Sinian-Cambrian, Carboniferous, Permian-Lower Triassic, Xujiahe Formation in Upper Triassic and Jurassic. From the exploration history, the exploration of Carboniferous "golden zone" lasted for more than 20 years, with proved geological reserves of natural gas of 2412×10^8 m^3; the exploration of reef-beach "golden zone" in Permian-Lower Triassic lasted for 16 years, with proved geological reserves of natural gas of 2922.3 $\times 10^8$ m^3; and the exploration of "golden zone" in Xujiahe Formation of Triassic lasted for 9 years, with proved geological reserves of natural gas of 7065.79 $\times 10$ m^3. The exploration of "golden zone" in Sinian-Cambrian has made no progress for decades since the discovery of Weiyuan Gasfield in 1960s. The large equipped gas field has been found in Moxi-Longwangmiao Formation since 2011, marked by the breakthrough of Well Gaoshi 1, with the proved geological reserves of natural gas of 4404×10^8 m^3, and the estimated reserves of more than trillion cubic meters. In addition, the new exploration "golden zones" are expected to be found in the strata of Qixia-Maokou Formation of Permian, Jialingjiang Formaion and Leikoupo Formation in Triassic where there is potential for new breakthroughs. It is proved that

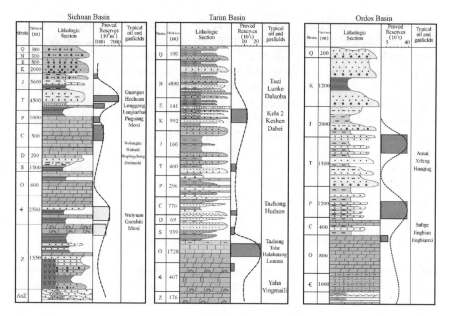

Fig. 4.43 Multiple exploration "golden zones" in Sichuan, Tarim and Ordos basins

the other exploration "golden zones" are also developed in the other superimposed petroliferous basins such as Tarim Basin, Ordor Basin et al.

(2)   Characteristics of Multi-Exploration "Golden Zones"

The theory of multi-exploration "golden zones" in the superimposed basin is a summary of the internal rules of accumulation according to the characteristics of the superimposed basin in China, and is different from the characteristics of multiple oil-bearing layers in the past. (1) The multi-stage characteristic of hydrocarbon source kitchen. The source rocks vertically in different strata and horizontally in different depressions are developed with the characteristics of multi-source and multi-period hydrocarbon supply, due to the differential subsident and multi-stage evolution in superimposed petroliferous basins. It is especially emphasized that, the gas source kitchens of late accumulation and effective accumulation are an important resource contributor to the breakthrough of the exploration forbidden zone in traditional "golden zones", due to the considerable number of liquid hydrocarbon remaining in the source rocks and the large amount of natural gas generated by further cracking in high-over mature stage. (2) The multi-stage development of reservoirs. Multiple sets of large-scale effective reservoirs are developed by the multi-cycle sedimentary structural evolution and various geological factors in superimposed basins. The constructive diagenesis of carbonate rocks runs through different stages of geological history, with the reservoir ranging from the middle-shallow layers to the middle-deep layers and even the ultra-deep layers. Multiple sets of reservoirs can be overlapped and connected on the plane, and muctiple layers developed large-scale vertically.

The large-scale multi-layer accumulation can be developed under the condition of abundant hydrocarbon sources and appropriate allocation of source and reservoir. (3) Multiple periods and late-stage effectiveness of accumulation. The multi-stage oil and gas accumulation is developed by the multi-stage hydrocarbon generation of source kitchen and multi-stage structural movements in or after several large migration periods of oil and gas. Among them, there are primary reservoirs developed by secondary migration from hydrocarbon source kitchen, adjusted reservoirs developed in new strata and new traps by adjustment in late-stage structure movements from the existing accumulations, the multiple accumulation occured in different regions by differential burial of the same hydrocarbon source kitchen, and multi-stage hydrocarbon generation transferred from oil ganeration to gas generation in the same hydrocarbon source kitchen with the increase of thermal evolution. As a result, the multi-stage accumulation can be developed by these factors.

(3)   The Significance of Multi-Exploration "Golden Zone"

The theory of multi-exploration "golden zones" in superimposed basins is a summary of the discovery rule of oil and gas in China in recent years, as well as the inheritance and development of the geology theory of oil and gas. It will have important guidance and reference significance for oil and gas exploration in superimposed basins, especially for deep oil and gas exploration in superimposed basins.

①   Multi-Exploration "Golden Zones" With Long Discovery History Leads to Multi-Peak Increase in Reserves

The multiple exploration "golden zones" are developed with the characteristics of multi-layer distribution of oil and gas caused by the multi-cycle structural sedimentary evolution of the superimposed basins. Recent exploration practice shows that, a new "golden zone" the reserve has will be found with the deepening of knowledge and the progress of engineering technology when a exploration "golden zone" is matured. For example, if based on the traditional petroleum geological theory, the oil and gas accumulation in paleo uplift of central Sichuan Basin is hard to form with the stage of high-over mature (Ro > 2.5%). While the understanding of deep oil and gas accumulation in paleouplift of the Middle Sichuan Basin is changed by the hydrocarbon generation theory of "successive gas generation" and "double peaks". The discovery of multiple sets of high-quality source rocks in the deep strata, the revaluation of dispersed liquid hydrocarbons (both inside and outside the source) and the cracking of paleo-oil reservoirs into gas, as well as the maturity of deep drilling technology, lead to the discovery of a trillion cubic meter gas field in Gaoshiti-Moxi area and push the natural gas exploration of Sichuan Basin to a new stage. The discovery process of oil and gas in Gaoshiti-Moxi area in the central Sichuan Basin shows that, the real underground conditions are gradually approached by the repeated of practice and recognition, due to the complex history of the accumulation in superimposed basin. This determines the complex exploration process and long discovery history of the superimposed basin, which has also the exploration potential.

② The Hydrocarbon Generation History is Complete, and The Resource Potential is Beyond Expectation

The hydrocarbon source kitchens composed of conventional source rocks and gas source kitchen of liquid hydrocarbon cracking are mainly distributed in large-scale superimposed petroliferous basins in China. The source kitchens of the conventional source rocks are mainly developed with two complete peaks of hydrocarbon generation, such as the "oil generation" and "gas generation", with sufficient evolution of source rocks and large amount of hydrocarbon generation. The gas source kitchens of liquid hydrocarbon cracking mainly include the late-stage cracking of the dispersed liquid hydrocarbon remaining in the source rocks, the "semi-accumulation and semi-dispersion" liquid hydrocarbon, and paleo oil reservoir. The contribution of paleoreservoir cracking to natural gas accumulation is considered in the early stage of resource evaluation, but the contribution of cracking gas of "semi-accumulation and semi-dispersion" liquid hydrocarbon and dispersed liquid hydrocarbon remaining in the source rocks are not considered. The breakthrough of Well Gucheng 6 in Gucheng of Tarim Basin confirms the reality of this kind of source kitchen exploration, which can make an important contribution to the deep large-scale accumulation of the superimposed basin. If considering the contribution of this part of liquid hydrocarbon cracking, the natural gas resources of the Lower Paleozoic in the Tarim Basin are 4.2×10^{12} m^3, 1.3 times higher than that of the third resource evaluation; the total gas resources of Sinian-Cambrian in Sichuan Basin are 12.8×10^{12} m^3, 2.4×10^{12} m^3 higher than that of the third resource evaluation. This shows the great exceedance of potential in deep-strata natural gas resources of China than expectations, as well as the better prospect of deep-strata discovery.

③ Economic Resources and Optimistic Exploration Prospect in the Deep Layers of the Superimposed Basin

The large-scale hydrocarbon supply can be developed by the hydrocarbon source kitchens of kerogen and liquid hydrocarbon cracking in deep layers of superimposed basins. Controlled by paleouplift, paleoslope, paleoplatform margin and multi-period successive fault zones, the multiple exploration "golden zones" of vertical superposition and horizontal combination are developed by multiple sets of large-scale effective reservoirs in the deep layers. The superimposed basins have economic resources and are worth exploring bacause of the wide rang of oil and gas and large scale of reserves in the deep layers even with the difference degree in the oil and gas enrichment in different structural parts. In recent years, the good prospect for deep oil and gas exploration is shown by the breakthroughs in deep-layer exploration of Tarim and Sichuan basins. In particular, the newly discovered gas field in Longwangmiao Formation of the central Sichuan Basin, is developed with not only a large scale of reserves with individual reserves of 4404×10^8 m^3 but also a high output in single well and good effect on production text. There are ten wells whose daily production is more than one million cubic meters, with a maximum open flow capacity of 1035×10^4 m^3 and a daily production capacity of 480×10^4 m^3. The exploration benefit cost of reserves discovery is less than 0.40 USD/bbl. The lower limit of 7000 m in industrial productivity has been exceeded in both the carbonate rocks of the platform area and Meso-Cenozoic clastic rocks in Kuqa foreland of Tarim Basin.

In a word, the exploration of "multiple golden zones" is a systematic understanding of the characteristics of multi-stage structural superimposition, multi-set source-reservoir configuration and multi-stage accumulation in superimposed basins of China. The accumulation factors of large oil and gas fields in deep carbonate reservoirs of China are summarized by the analysis of formation mechanism guided by the "multiple golden zones" theory. As the key targets of multi "golden zones" exploration, the favourable strata for the oil and gas accumulation in different oil and gas reservoirs are controlled by the paleouplift, paleoslope, paleoplatform margin and multi-stage successive fault belts, in the complex geological background of vertical superposition of multiple sets of marine (continental) sedimentary sequences and horizontal compound of multi-phase multi-type sedimentary sequence. It is predicted that, the deep and ultra-deep oil and gas fields will make great contributions to the development of oil and gas industry in China as well as the oil and gas prospects are worth looking forward to with the deepening of geological understanding and the progress of engineering technology.

References

Chengzao, Jia, He. Dengfa, Shi Xin, et al. 2006. Characteristics of China's oil and gas pool formation in latest geological history[J]. *Science in China* 36 (5): 412–420.

Chengzao, Jia, Li Benliang, Zhang Xingyang et al. 2007. Formation and evolution of marine basin in China[J]. *Chinese Science Bulletin*, 52 (supplement I), 1–8.

Chuanbo, Shen, Xu. Mei Lianfu, Zhenping, , et al. 2007. Fission track evidence of uplift in middle-late Cenozoic of Dabashan[J]. *Acta Petrologica Sinica* 23 (11): 2901–2910.

Claypool, G.E., and E.A. Mancini. 1989. Geochemical relationships of petroleum in Mesozoic reservoirs to carbonate source rocks of Jurassic Smackover Formation, Southwest Alabama[J]. *AAPG Bulletin* 73 (10): 904–924.

Du Jinhu, Hu., Zhang Yijie Suyun, et al. 2013. Implications from typical petroleum exploration cases[J]. *Acta Petrolei Sinica* 34 (5): 809–819.

Guohui, Li., Li. Xiang, and Yang Xinan. 2000. Controlling factors of Sinian gas pools in Caledonian Paleouplift, Sichuan Basin[J]. *Oil & Gas Geology* 21 (1): 80–83.

Hui, Tian, Wang Zhaoming, Xiao Zhongyao, et al. 2006. Oil cracking to gases: kinetic modeling and geological significance[J]. *Chinese Science Bulletin* 51 (15): 1821–1827.

Jinxing, Dai, Zou Caineng, Tao Shizhen, et al. 2007. Forming conditions and main controlling factors of large gas fields in China[J]. *Natural Gas Geoscience* 18 (4): 473–484.

Richardson, N.J., A.L. Densmore, D. Seward, et al. 2008. Extraordinary denudation in the Sichuan Basin: Insights from low-temperature thermochronology adjacent to the eastern margin of the Tibetan Plateau[J]. *Journal of Geophysical Research* 113: 1–23.

Shuichang, Zhang, and Zhu Guangyou. 2006. Gas accumulation characteristics and exploration potential of marine sediments in the Sichuan Basin. *Acta Petrolei Sinica* 27 (5): 1–8.

Shuichang, Zhang, Zhang Baomin, Li. Benliang, et al. 2011. History of hydrocarbon accumulations spanning important tectonic phases in marine sedimentary basins of China: Taking the Tarim Basin as an example[J]. *Petroleum Exploration and Development* 38 (1): 1–15.

Shuyuan, Shi, Liu Wei, Jiang Hua, et al. 2015. Characteristics of Paleocene fault-fracture system and their relationship with Ordivician paleokarst reservoirs in Halahatang area, north Tarim Basin[J]. *Journal of Central South University (science and Technology)* 12: 4568–4577.

Wenzhi, Zhao, Wang Zecheng, and Wang Yigang. 2006a. Formation mechanism of highly effective gas pools in the Feixianguan Formation in the NE Sichuan Basin[J]. *Geological Review* 52 (5): 708–718.

Wenzhi, Zhao, Wang Zhaoyun, Wang Hongjun, et al. 2006b. Cracking conditions of oils existing in different modes of occurrence and forward and backward inference of gas source rock kitchen of oil cracking type[J]. *Geology in China* 33 (5): 952–965.

Wenzhi, Zhao, Wang Zhaoyun, Wang Hongjun, et al. 2011. Further discussion on the connotation and significance of the natural gas relaying generation model from organic materials[J]. *Petroleum Exploration and Development* 38 (2): 129–135.

Zecheng, Wang, Jiang Hua, Wang Tongshan, et al. 2014. Paleo-geomorphology formed during Tongwan tectonization in the Sichuan Basin and its significance for hydrocarbon accumulation[J]. *Petroleum Exploration & Development* 41 (3): 305–312.

Zhang Shuichang, Hu., Mi Jingkui Guoyi, et al. 2013. Tine-limit and yield of natural gas generation from different origins and their effects on forecast of deep oil and gas resources[J]. *Acta Petrolei Sinica* 34 (Supplement 1): 41–50.

Zhaoyun, Wang, Zhao Wenzhi, and Wang Yunpeng. 2004. China's marine carbonate source rock evaluation research [J]. *Progress in Natural Science* 14 (11): 1236–1243.

Zhengwu, Guo, Deng Kangling, and Han Yonghui. 1996. *Formation and evolution of the Sichuan Basin[M]*. Beijing: Geological Publishing House.

Chapter 5
Prospect for Exploration of Deep Oil and Gas Fields Onshore in China

The exploration of deep oil and gas will play an important role in the sustainable development of oil and gas resources in China, and the deep oil and gas resources will become an important strategic area of energy security with the rapid economic growth and relative shortage of oil and gas supply. The deep oil and gas resources onshore are concentrated in three major fields such as deep carbonate reservoir, deep clastic reservoirs and deep volcanic reservoirs, and a number of deep oil and gas fields have been discovered, such as Puguang Gasfield, Yuanba Gasfield, Anyue Gasfield, Keshen Gasfield, Tahe Oilfield, Shunbei Oilfield and so on. The deep reservoir onshore will be one of the major fields of oil and gas development in the 13th Five-Year Plan and even in the future, with the further improvement of development technology. According to the existing accumulation geological analysis, three major favorable exploration fields for deep oil and gas are developed, including gypsum-carbonate combination, middle-late Proterozoic and deep multi-strata thrust belt. The resource potential evaluation, favorable zone evaluation and exploration targets of deep oil and gas can be developed, using the evaluation technology in deep-layer exploration, which is formed by determining the parameters, standards and method flow of deep favorable zone evaluation with the arrangement of oil and gas reservoir data and the compilation of software.

5.1 Prospect of Deep Oil and Gas Exploration Onshore in China

The exploration in deep oil and gas is the key field of international oil and gas exploration and discovery and also of increasing reserves and production during the 13th Five-Year Plan in China. The total amount of remaining resources in deep oil and gas exploration is huge, with low proportion of proved reserves. Especially the deep carbonate reservoirs have more remaining resources than that of any other lithology with huge exploration potential. As a result, based on the research of oil

© Petroleum Industry Press 2021
S. Hu and T. Wang, *Deep-Buried Large Hydrocarbon Fields Onshore China: Formation and Distribution*, https://doi.org/10.1007/978-981-16-2285-4_5

and gas accumulation, the realistic zones of deep oil and gas exploration in the 13th Five-Year Plan are selected, including five carbonate zones, seven clastic zones and four volcanic zones.

5.1.1   Hydrocarbon Potential in Deep Strata

Nowadays, the oil and gas exploration is expanded from middle-shallow strata to deep strata and ultra-deep strata in terms of exploration depth and from conventional oil and gas to conventional and unconventional oil and gas in terms of resource types and fields. As a result, the exploration in deep strata with buried depth greater than 4500 m has become an important field for increasing reserves and production of oil and gas in the world. At present, the global discovered deep-strata reserves account for 40% of the total reserves, with the increased oil production from 1.21×10^8 t to 1.5×10^8 t and the increased deep natural gas production from 1054×10^8 m^3 to 1400×10^8 m^3 in the past five years. The oil and gas exploration potential of ultra deep and ancient formations is great, and a series of major breakthroughs have been made in the three major fields of deep exploration vis-à-vis carbonate rocks, clastic rocks and volcanic rocks. Taking large-scale gas fields of Kuqa and Yuanba as examples, the reserves of natural gas in ultra-deep (>7000 m) exceeds trillion cubic meters; the major breakthroughs have been made in oil exploration of ultra-deep strata represented by transition zone of Tabei-Tazhong (7200–7500 m), Kuqa-Keshen-Dabei (6900–7500 m) and so on; the reserves of natural gas in Proterozoic-Cambrian in Anyue Gasfield in Sichuan Basin exceeds 1.56×10^{12} m^3, represented by oversized. According to the new results of national resource evaluation, the remaining deep oil and gas resources are rich, the resource proved rate is low (5.9% for oil and 6.1% for gas), and the potential of finding new reserves is great. The remaining oil resource is 304×10^8 t, with proved oil reserve is 18×10^8 t, and the remaining resource is 286×10^8 t; the remaining natural gas resource is 26.88×10^{12} m^3, with proved reserve of 1.64×10^{12} m^3, and the remaining resource is 25.24×10^{12} m^3. According to the statistics in Table 5.1, the proportion of remaining resources in deep carbonate reservoir is relatively high, and the exploration potential is great. The remaining reserve of crude oil in deep carbonate reservoir is 93.8×10^8 t, accounting for 57.5%, and natural gas is higher, with the remaining amount of 11.5×10^{12} m^3, accounting for 76.8%; the remaining reserve of deep clastic reservoir is middle, and the proportion of remaining reserves in volcanic reservoir and metamorphic reservoir is the lowest.

Table 5.1 Statistics of resources potential of deep oil and gas

Fields	Oil resources (10^8 t)				Natural gas resources (10^{12} m^3)			
	Resource extent	Proved reserves	Remaining resources	Remaining resource rate (%)	Resource extent	Proved reserves	Remaining resources	Remaining resource rate (%)
Deep carbonate reservoir	98.7	4.9	93.8	57.5	11.8	0.3	11.5	76.8
Deep clastic reservoir	48.6	2.1	46.5	28.5	3.17	1.1	2.07	13.8
Deep volcanic reservoir	22.1	4.8	17.3	10.6	2	0.6	1.40	9.4
Deep metamorphic reservoir	7.0	1.5	5.5	3.4				
Summation	176.4	13.3	163.1	100.0	16.97	2.0	14.97	100.0

5.1.2 Prospect of Deep Oil and Gas Exploration

(1)   Deep Carbonate Reservoir

The exploration of marine carbonate reservoir has become the focus of oil and gas exploration in the 13th Five-Year Plan. Based on the research of oil and gas accumulation, five practical zones (Table 5.2) are selected, including Halahatang and Tazhong in Tarim Basin, Lower Paleozoic in Ordos Basin, Longwangmiao Formation and Dengying Formation in the Central of Sichuan Basin. The increased reserves are mainly natural gas (7500 $\times$ 10^8 m^3), accounting for 27% of the planning. Five succeeding zones are selected, including the northern and southern horst zones in Gucheng, the under-salt zone of Cambrian in Tazhong, the under-salt zone of Lower Paleozoic in Jingbian, Sinian-Cambrian strata in the southern and central of Sichuan Basin, and Maokou -Qixia Formation in the central of Sichuan Basin. Six preparation zones are selected, including low uplift in Manxi, the under-salt zone of Cambrian in Bachu, rift trough of Middle-Late Proterozoic and Ordovician in the eastern of Ordos Basin, reef-beach of Permian–Triassic in the northeast of Sichuan Basin and Sinian-Cambrian in the eastern of Sichuan Basin.

(2)   Deep Clastic Reservoir

During the the 13th Five-Year Plan, the exploration fields of deep clastic reservoirs (Table 5.3) are developed, including deep natural gas in Kuqa area, natural gas in Xujiahe Formation in Sichuan Basin, and lithologic strata in the hinterland of Junggar Basin, with a total area of 9.5 $\times$ 10^4 km^2 which have oil resources of 8 $\times$ 10^8 t and natural gas of (4.5–5.2) $\times$ 10^{12} m^3. Seven major fields of oil and gas succeeding exploration are developed, including deep clastic reservoir in Bohai Bay, deep marine sandstone reservoir in Tarim Basin, natural gas of deep tight sandstone reservoir in Junggar Basin, oil and gas in deep strata of southwestern of Tarim Basin, tight gas in Taibei Depression of Turpan-Hami Basin, tight oil in Santanghu Basin and deep tight gas in Songliao Basin, with total exploration area of about 34 $\times$ 10^4 km^2, oil resources of (29.8–42.3) $\times$ 10^8 t and natural gas of (5.5–9.0) $\times$ 10^{12} m^3.

(3)   Deep Volcanic Reservoir

The deep volcanic rocks onshore are developed with the area of 36 $\times$ 10^4 km^2, with resources of about 50 $\times$ 10^8 t oil equivalent, the proven rate of 10%, which have great potential. The realistic fields (Table 5.4) include Carboniferous and Permian of Junggar Basin, Jurassic and Cretaceous of Songliao Basin, Carboniferous and Permian of Santanghu Basin and Jurassic and Paleogene of Bohai Bay Basin, with exploration area of 14 $\times$ 10^4 km^2, remaining oil resources of 11.3 $\times$ 10^8 t and remaining gas resources of 2.97 $\times$ 10^{12} m^3. The succeeding fields include Permian of Tarim Basin, Carboniferous and Permian of Turpan-Hami Basin and Permian of Sichuan Basin, with exploration area of 17.5 $\times$ 10^4 km^2, oil resources of 6.5 $\times$ 10^8 t and gas resources of 1 $\times$ 10^{12} m^3.

Table 5.2 Statistics of oil and gas exploration prospect in carbonate rocks of the 13th five-year plan

Number	Basin	Zone	Reverse (Oil 10^8 t) (Gas 10^8 m^3)	Basic information			Potential of increasing reserves in the next five years	Types of field
				Remaining controlled and predicted reserve	Proved area	Provend rate		
1	Tarim Basin	Halahatang	10 (Oil)	0.93×10^8 t (Control)	2.2×10^8 t	22%	1.5×10^8 t	Realistic
2		Tazhong	12,000 (Gas)	Controlled and predicted reserves of 2886×10^8 m^3	4012×10^8 m^3	33%	1500×10^8 m^3	Realistic
3		Tazhong (under salt in Cambrian)	4000 (Gas)	Obtained from Well Zhongshen 1C, with favorable area of 2100 km^2				Superseding
4		Northern and southern horsts in Gucheng area	5100 (Gas)	Obtained from Well Gucheng 6, with favorable area of 4300 km^2				Superseding
5		Low uplift in Manxi		Uplift in early stage, with favorable preservation in late stage				Preparation
6		Under salt in Cambrian in Bachu		Inherited paleouplift, with good accumulation conditions				Preparation
7	Ordos Basin	Lower Paleozoic	25,000 (Gas)	Controlled and predicted reserves of 1533×10^8 m^3	6547×10^8 m^3	26%	1500×10^8 m^3	Realistic

(continued)

Table 5.2 (continued)

Number	Basin	Zone	Reverse (Oil 10^8 t) (Gas 10^8 m^3)	Basic information				Potential of increasing reserves in the next five years	Types of field
				Remaining controlled and predicted reserve	Proved area	Provend rate			
8		Jingbian (under salt in Lower Paleozoic)	1000 (Gas)	Obtained from Well Tao 38 and Jintan 1, with favorable area of 1115 km^2					Superseding
9		Eastern Ordovician		Breakthrough in exploration of middle combination in Ordovician on the eastern paleouplift					Preparation
10		Rift trough in Middle-Late Proterozoic		Multiple sets of source rocks with rich organic carbon are developed, and large-scale reservoirs with microbial diagenesis are developed					Preparation
11	Sichuan Basin	Longwangmiao Formation in the central of Sichuan Basin	42,600 (Gas)	Predicted reserve of 528×10^8 m^3	4404×10^8 m^3	10%		1500×10^8 m^3	Realistic

(continued)

Table 5.2 (continued)

Number	Basin	Zone	Reverse (Oil 10^8 t) (Gas 10^8 m^3)	Basic information				Potential of increasing reserves in the next five years	Types of field
				Remaining controlled and predicted reserve	Proved area	Provend rate			
12		Dengying Formation in the central of Sichuan Basin	24,400 (Gas)	Controlled and predicted reserves of 9575×10^8 m^3				3000×10^8 m^3	Realistic
13		Southern-central of Sichuan Basin (Sinian–Cambrian)	5000 (Gas)	Obtained from Well Heshen 1 and Nanchong 1, with favorable area of 1860 km^2					Superseding
14		Central of Sichuan Basin (Qixia Formation-Maokou Formation)	5000 (Gas)	Obtained from Well Nanchong 1, with favorable area of 5000 km^2					Superseding
15		Sinian–Cambrian in eastern of Sichuan Basin		Source rocks in Cambrian are developed, and dolomite reservoirs of beach facies are widely distributed					Preparation
16		Reef beach in Permian–Triassic of northeast of Sichuan Basin		The accumulation conditions of Longgang, Poxi are similar with that of adjacent Puguang and Tieshanpo					Preparation

Table 5.3 Statistics of oil and gas exploration prospects of clastic reserovirs in the 13th five-year plan

Types of field	Exploration field	Favorable conditions	Exploration area (10^4 km^2)	Resource scale
Realistic	Deep natural gas in Kuqa	Favorable source-reservoir combination and stable caprock of gypsum salt	2.3	$(2.69–3.42) \times 10^{12}$ m^3
	Natural gas in Xujiahe Formation in Sichuan Basin	The "sandwich" structure is formed by hydrocarbon generating center of coal measures and favorable reservoir sand bodies ructure	3.8	1.8×10^{12} m^3
	Lithologic strata in the hinterland of Junggar Basin	Multiple hydrocarbon generating depressions, large-scale sedimentary system and favorable migration and accumulation conditions	3.4	8×10^8 t
Superseding	Deep clastic reservoirs in Bohai Bay Basin	Oil and gas accumulation is developed by over-pressure filling, which is formed by the close contact with the main source rocks in large area, with the development of depressions in Es1-Es4, submarine fan on the far side of the slope, braided river (fan) delta front sand body	6.5	Oil $(15–20) \times 10^8$ t Gas $(1.0–2.0) \times 10^{12}$ m^3

(continued)

Table 5.3 (continued)

Types of field	Exploration field	Favorable conditions	Exploration area (10^4 km^2)	Resource scale
	Marine sandy reservoirs in Tarim Basin	Favorable conditions for large-scale oil and gas accumulation are developed by good connectivity of sand body, the background of paleo-uplift slope, development of passage system such as fault and unconformity, as well as good preservation conditions	20	Oil $(8.5–14) \times 10^8$ t Gas $(0.5–1) \times 10^{12}$ m^3
	Natural gas in deep tight sand reservoirs in Jungar Basin	Favorable combination of source and reservoir with large-scale distribution of tight sandstones	2	$(0.8–1.2) \times 10^{12}$ m^3
	Deep oil and gas in the Southwest of Tarim basin	Three sets of major source rocks are developed: C, P and J; three sets of oil-bearing strata are developed: Neogene, Paleogene and Cretaceous	3	Oil 3.3×10^8 t Gas 1.3×10^{12} m^3
	Tight gas in Taibei Depression of Turpan-Hami Basin	Large-scale tight sandstone gas reservoir is developed by large-area tight sandbody in mature source rocks of coal measures of the Turpan-Hami Basin	1	$(0.6–0.9) \times 10^{12}$ m^3

(continued)

Table 5.3 (continued)

Types of field	Exploration field	Favorable conditions	Exploration area (10^4 km^2)	Resource scale
	Tight oil in Santanghu Basin	Favorable combination of source and reservoir with large-scale distribution of tight sandstones	0.4	$(3–5) \times 10^8$ t
	Deep tight gas in Songliao Basin	Favorable source rocks are developed in the multiple-fault depression, as well as large-scale sandbodies, with good preservation conditions	1.3	$(1.32–2.53) \times 10^{12}$ m^3

5.2 Favorable Exploration Areas of Deep Oil and Gas

The onshore deep layers in China have geological conditions and resource base to form large oil and gas fields, with great exploration potential. Three types of deep oil and gas exploration areas are worthy of attention in terms of current geological research. Nowadays, the combination of gypsum-salt rock and carbonate rock is developed as an important field of oil and gas exploration in the world, and it is widely developed with relatively low exploration degree in the three major marine basins in China, which will become an important succeeding field once explored. Exploration at home and abroad has confirmed that large-scale oil and gas fields can be found in the Middle-Late Proterozoic, which has favourable oil and gas geological conditions and great accumulation potential. The deep multi-strata system of thrust belt is worthy of attention. Generally speaking, the oil and gas accumulation is mainly developed by the scale of reservoir and the effectiveness of trap, but not the source rocks and structures.

5.2.1 Combination of Gypsum-Salt Rock and Carbonate Rock

Globally, the combination of gypsum-salt rock and carbonate rock is an important field to discover oil and gas, accounting for 65% of the total reserves of marine carbonate rocks. There are 115 gypsum-salt sedimentary basins in the world (Fig. 5.1), and 97 basins are related to oil and gas, 66 of which are rich in oil and gas. As of 2012, oil recoverable reserves of 665×10^8 t and natural gas recoverable

Table 5.4 Statistics of oil and gas exploration prospects of volcanic reserovirs in the 13th five-year plan

Types of field	Exploration field	Favorable conditions	Exploration area (10^4 km^2)	Resource scale	Proved reserves
Realistic	Jungar Basin C, P	Effective combination of weathering crust reservoir, volcanic rocks and source rocks and 8 residual fault depressions	6	Oil 9.6 × 10^8 t Gas 0.8 × 10^{12} m^3	Oil 3.3 × 10^8 t Gas 1266 × 10^{12} m^3
	Songliao Basin J, K	Good gas generation conditions are developed in 26 depressions with the superimposed development of volcanic rocks	5	Gas 2 × 10^{12} m^3	Oil 196 × 10^4 t Gas 4560 × 10^8 m^3
	Santanghu Basin C, P	The weathering crust and volcanic body of volcanic rocks are well developed and well distributed with source rock	1	Oil 3.5 × 10^8 t	Oil 4370 × 10^4 t
	Bohai Bay Basin J, E	Volcanic rocks are developed in the source rocks, which are favorable for oil and gas accumulation	2	Oil 3 × 10^8 t Gas 0.8 × 10^{12} m^3	Oil 9971 × 10^4 t
Superseding	Tarim Basin P	Good combination of source rock and reservoir is developed by the weathering crust and rock body of volcanic rocks and the underlying source rocks	13	Oil 3 × 10^8 t	

(continued)

Table 5.4 (continued)

Types of field	Exploration field	Favorable conditions	Exploration area (10^4 km^2)	Resource scale	Proved reserves
	Turpan-Hami Basin C, P	Volcanic weathering crust and rock body are adjacent to source rocks	2	Oil 3.5 × 10^8 t Gas 0.55 × 10^{12} m^3	
	Sichuan Basin P	Good configuration of volcanic rocks and underlying source rocks	2.5	Gas 0.45 × 10^{12} m^3	30 × 10^8 m^3

Fig. 5.1 Distribution of global gypsum-salt sedimentary basins

reserves of 103 × 10^{12} m^3 have been found. Such as oil of 1221.32 × 10^6 bbl and gas of 6575.2 × 10^8 ft^3 found in the Devonian of Pripyat Basin, oil of 49.25 × 10^6 bbl and gas of 2558.0 × 10^8 ft^3 obtained in Cambrian of Alborz and Sarajeh Oilfields in Kum Basin, oil of 855.75 × 10^6 bbl and gas of 1925.8 × 10^8 ft^3, oil of 611.77 × 10^6 bbl and gas of 81,191.3 × 10^8 ft^3 obtained respectively in the Alkhlata and Amin structural belts of Oman Basin.

The combinations of gypsum-salt rocks and carbonate rocks are widely developed in the three major marine basins in China. Once a breakthrough is made, it will become an important succeeding field for oil and gas exploration although the exploration level is relatively low at present. First, the combinations of gypsum-salt rocks and carbonate rocks are widely distributed in China, with many important discoveries made in recent years. The combinations of gypsum-salt rocks and carbonate rocks are developed in all the three craton basins, especially in the strata of Cambrian-Ordovician, Carboniferous-Permian and Middle-Lower Triassic. Recently, large-scale reserves have been found in Longwangmiao Formation of Sichuan Basin, and important discoveries have been made in Well Zhongshen 1 of Tarim Basin and Majiagou Formation of Ordos Basin, with favorable area of 1×10^4 km^2 through a preliminary evaluation. Secondly, the accumulation potential in the combination of gypsum-salt rocks and carbonate rocks is great, with the match of accumulation factors. The high-quality source rocks are developed in the combinations of gypsum-salt rocks and carbonate rocks, with the increased hydrocarbon generation potential of organic matter. The dolomite reservoirs with large-scale distribution can be developed in the grainbeach associated with gypsum-salt rocks. The structural traps can be developed in up-salt and under-salt strata, as gypsum-salt rocks can be used as good caprocks and detachment strata. At the same time, the up-salt traps and under-salt traps, as well as the good migration channels such as faults and dislocation of salt bodies are formed by the discordant deformation between up-salt and under-salt strata under the tectonic compression environment.

Two types of accumulation combination of up-salt and under-salt may be developed in the Cambrian of the eastern and southern of Sichuan Basin. The new structure model of deep strata in the eastern of Sichuan Basin is built that, up-salt ramp thrust structures and under-salt large-scale structures are developed according to the viewpoint of seismic section and balanced section (Fig. 5.2). Based on the research of accumulation conditions, two accumulation models of under-salt and up-salt may be developed in the Cambrian in the eastern of Sichuan Basin. The target layers

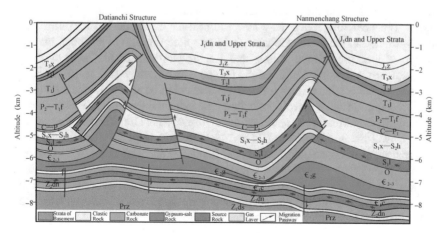

Fig. 5.2 Geological profile of the development of the under-salt and up-salt gypsum-salt reservoirs of Cambrian in the eastern of Sichuan Basin

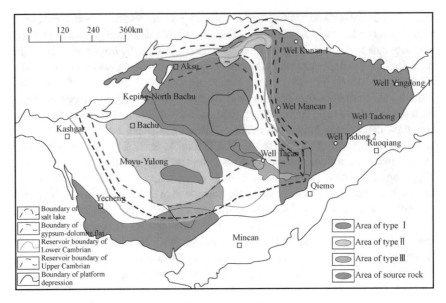

Fig. 5.3 Under-salt exploration and evaluation in Tarim Basin

of the up-salt accumulation combination are Xixiangchi Formation, Longwangmiao Formation and Ordovician, with the area of 1370 km^2. The target layers of the under-salt accumulation combination are Dengying Formation, with the area of 488 km^2. And the exploration in target layers of the under-salt accumulation is risky, as the trap credibility and reservoir prediction are difficult with the poor quality of the data and deep buried depth.

The favorable exploration targets are developed in under-salt dolomite reservoirs of Cambrian in Tarim Basin and grain beach in annulus of salt depression of Ordos Basin. The favorable zones (Fig. 5.3) within a depth of 8000 m in Tarim Basin are distributed in the Eastern Bachu-Tazhong, platform margin of Tabei, platform margin of Gucheng, Moyu-Yulong and Keping-Northern Bachu, with the I-type area of nearly 1.7×10^4 km^2, and the II-type area of nearly 1.15×10^4 km^2. The exploration risk is relatively high, due to the deeper buried depth, and the difficulty of target recognition and reservoir prediction. The industrial gas flow has been obtained in many wells of Member 5_{7-9} of Majiagou Formation in Ordos Basin, such as Well Tao 39, Well Jintan 1, Well Tong 74 and Well Tong 75 (Fig. 5.4), and the favorable exploration target layers are Member 5_7, Member 5_9 and Member 4. It is confirmed that the grain beach around the salt depression (with an area of about $(1.5–2.0) \times 10^4$ km^2) and the grain beach under the salt (with an area of about $(1.0–1.5) \times 10^4$ km^2) have good exploration prospects and they also have certain risks with the scale of the hydrocarbon source kitchen under the salt.

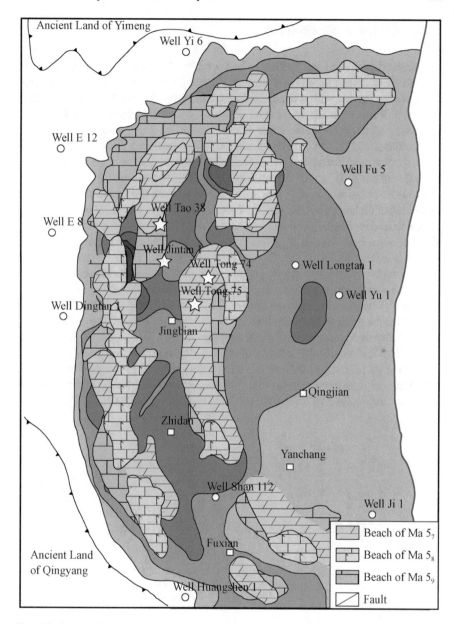

Fig. 5.4 Superposition of grain beach in Member 5_{7-9} and *source* rocks in Member 5_{5-10} of Majiagou Formation of Ordos Basin

5.2.2   Middle-Late Proterozoic

The large-scale oil and gas fields can be found in Middle-Late Proterozoic by the exploration practice at home and abroad (Fig. 5.5). 64 oil and gas fields have been discovered in Russia, with proved and controlled reserves of 22.36×10^8 t oil equivalent by the end of 2005. More than 90% of Oman's oil production comes from the source rocks in Neoproterozoic-Lower Cambrian. The geological reserves in Neoproterozoic-Cambrian of Baghavala Oilfield of India are 6.28×10^8 bbl. The gas-bearing area of Gaoshiti-Moxi of Sinian in China is 7500 km^2, the proved natural gas reserves are 2201×10^8 m^3, and the controlled reserves are 2038×10^8 m^3.

The accumulation potential has been developed with the favourable geological conditions of oil and gas in the Middle-Late Proterozoic in China. The ancient high-quality source rocks are developed in Middle-Late Proterozoic, due to the prosperity of lower organisms caused by climate environment, interglacial period, volcanic material, radioactivity and other factors. At the same time, the large-scale reservoirs can be formed by constructive diagenesis of the microbiolites. Microbiolites such as thrombolite, foam layer, granular dolomite are developed in the Member 4 of Dengying Formation in Proterozoic (Fig. 5.6) in Well Gaoke 1, Sichuan Basin of in, and several types of pores such as the microbial framework (dissolution) pores, intergranular dissolved pores and dissolved pores are also developed with the porosity of 2–10% (3.3% on average), permeability of 1–10 mD (2.26 mD on average).

Large-scale rift troughs are developed in Proterozoic in three major craton basins, with great exploration potential. The relatively complete strata in Proterozoic-Cambrian are retained in three craton basins of China, North China, Yangtze and

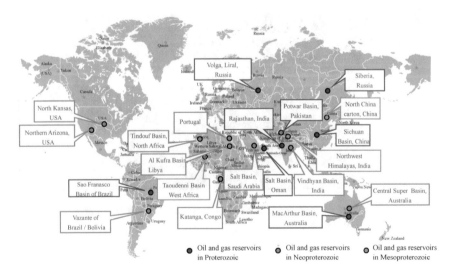

Fig. 5.5 Exploration and discovery of oil and gas in Middle-Late Proterozoic in the world

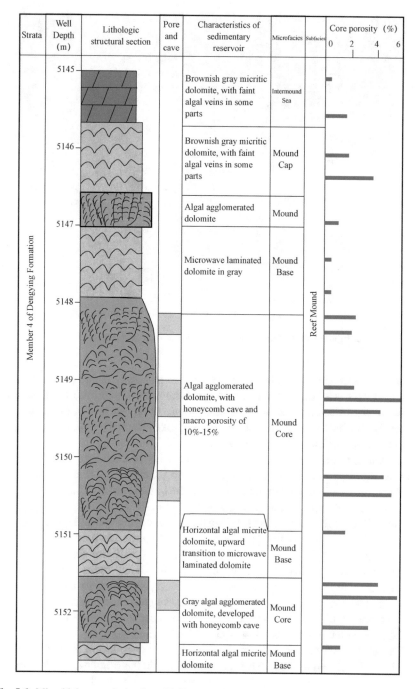

Fig. 5.6 Microbial reservoir developed in Proterozoic of Well Gaoke 1 in Sichuan Basin

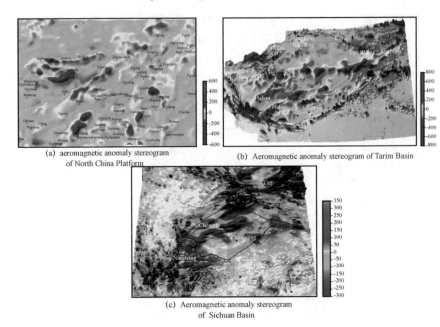

(a) aeromagnetic anomaly stereogram
of North China Platform

(b) Aeromagnetic anomaly stereogram of Tarim Basin

(c) Aeromagnetic anomaly stereogram
of Sichuan Basin

Fig. 5.7 The distribution map of aeromagnetic anomaly in three major cratons

Tarim. The late-stage development of basins, the paleogeographic pattern of sedimentary facies, the scale of source rocks and the reservoir-caprock combinations, as well as the petroleum geological characteristics of the Meso-Neoproterozoic are obviously effected by the formation, evolution and development scale of depressions in craton basins of Meso-Neoproterozoic.

The whole Ordos Basin (Fig. 5.7a) is formed with the extension from NE to SE, with the development of NE rift troughs in deep strata of Chancheng System. The Gansu-Shaanxi Rift Trough in the northern of the basin extends to the northeast and may be connected with the Xingmeng Rift Trough in the north margin. The Shanxi-Shaanxi Rift Trough in the southern basin extends eastward to Qinshui Basin, and then connects with Yanliao Rift Trough. The large-scale source rocks may be developed in Changcheng System in Ordos Basin, and distribute in the rift trough. The secondary accumulation combination may be developed, with source rocks of Changcheng-Jixian System in Proterozoic—Cambrian, and overlying reservoirs and caprocks of Paleozoic and Mesozoic. The major part of rift trough should be considered as the next exploration target. In the area developed with deep and large faults, natural gas can be transported to the shallow strata of Paleozoic and Mesozoic.

The residual source rocks should be developed in deep layers of Proterozoic in Bohai Bay Basin. More than 70 oil seeps are found in the Middle-Late Proterozoic in Yanliao area. Through the coring from the Proterozoic in Chengde area, it can be seen that the cores of Hongshuizhuang Formation and Chuanlinggou Formation are oil-bearing and full of crude oil, which should come from the Proterozoic strata itself. According to comprehensive analysis, the source rocks in Proterozoic may be widely developed in Yanliao area, with moderate degree of thermal evolution, and in the early stage of oil generation and gas generation. The favourable areas for primary oil and gas reservoirs of Proterozoic should be in the areas covered by strata of Paleogene and Neogene and well preserved, such as Jibei Depression and Northern of Jizhong depression.

A rift of near E-W trend in craton basin may be developed in the northern of Tarim Basin (Fig. 5.7b), while two rifts of S-E trend in craton basin are developed in the south. At present, there is no wells drilled in this set of source rocks in the basin. Only black shale has been found in the outcrop section of Middle-Lower Tereeken Formation in Yaerdangshan, with a thickness of 326.59 m and an average TOC of 2.96%. It is speculated that the Proterozoic source rocks are likely to be developed in the strata of Nanhua-Sinian in Tarim Basin, which are the potential source rocks.

The aeromagnetic anomaly in Sichuan Basin shows as band in NE (Fig. 5.7c). It is speculated that large-scale craton riftings in Nanhua system are developed in Sichuan Basin. The basin experienced the volcanic activities in the early rift period, and sedimentary filling in the rift period which is filled with clastic rocks with large thickness. Outcrops show that high-quality source rocks are developed in Doushantuo Formation of Sinian and Datangpo Formation of Nanhua System. The constructively modified microbialites in Sinian and grain beach in carbonate rocks in Cambrian and above strata can be developed as reservoirs. And three types of accumulation combinations are possibly developed, such as lower-generation and upper-storage, new-generation and old-storage, upper-generation and lower-storage (Fig. 5.8). The platform margin uplift belts distributed around the rift trough which controls the development of source kitchens is the favorable zones for oil and gas accumulation.

In summary, large-scale rifts in craton basins are developed in the Middle-Late Proterozoic of the three major craton areas in North China, Yangtze and Tarim in China, which control a large scale of hydrocarbon source kitchens with great potential

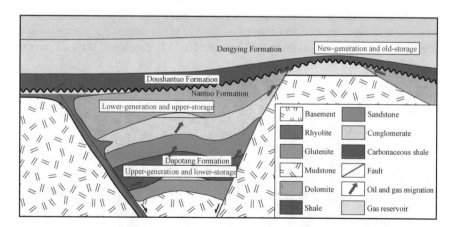

Fig. 5.8 Accumulation combination of large-scale rift trough in Neoproterozoic of Yangtze Craton

of thermal cracking into gas in the high-over mature stage. The large-scale effective reservoirs can be formed through multi-stage constructive transformation of microbial carbonate rocks and granular dolomites in Sinian, Cambrian and above strata with realistic exploration value and exploration potential. In the next step, the new breakthrough in oil and gas exploration of Proterozoic-Cambrian is expected to be achieved by evaluating and selecting favorable exploration areas through the research of sequence correlation by meticulous depiction of distribution of the rifts in craton.

5.2.3 Deep Multi-strata System of Thrust Belt

For the complex structure in the western of Sichuan Basin, the key problems are the reservoir scale and the trap effectiveness, rather than the source rocks and structures. In this area, 3 large-scale gas source kitchens (Qiongzhusi Formation, Upper Permian and Xujiahe Formation), 4–7 rows of thrust structures (Fig. 5.9), 18 gas reservoirs and more than 30 gas-bearing structures are found. Among them, many high-yield gas flow wells have been discoveried, such as Well Shuangtan 1 and Well Chuanke 1. The distance between Well Kuang 2 and Well Kuang 3 is 9 km. The reservoir thickness of Qixia Formation of Well Kuang 2 is 44 m, while Well Kuang 3 has no reservoiso. Water is produced in many wells of Leikoupo Formation in the southwest of Sichuan Basin, such as Well Wu 1. There are also differences in reservoir target strata and trap integrity of thrust belts in the northwest and southwest of Sichuan Basin.

Large-scale reservoirs are developed in the multiple marine strata in the northwest of Sichuan Basin and the key is reservoir evaluation and trap implementation. Seven sets of gas-bearing strata (Ordovician, Qixia Formation, Maokou Formation, Wujiaping Foramtion, Changxing Formation, Feixianguan Formation and Leikoupo Formation) and six sets of high-energy facies belts (Sinian, Cambrian, Qixia Formation, Changxing Formation, Feixianguan Formation and Leikoupo Formation) have been found in northwest of Sichuan, of which Qixia Formation deserves attention. The Qixia Formation in northwest of Sichuan Basin is developed with an area of

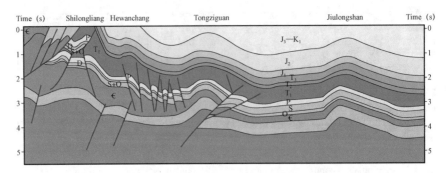

Fig. 5.9 Geological section of deep structure in thrust belt in the northwest of Sichuan Basin

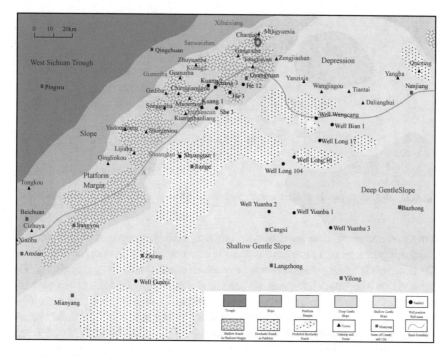

Fig. 5.10 Sedimentary and lithofacies palaeogeography of Qixia Formation of Permian in the northwest of Sichuan Basin

5000–8000 km^2 (Fig. 5.10), a thickness of 20–100 m and a reservoir porosity of 2–7.5%. The sedimentary facies are platform margin beach and inner platform beach. The reservoirs of platform margin are developed in Nianziba, with a thickness of 100 m, and the reservoirs within the platform are developed in Well Shuangtan 1, with a thickness of about 20 m. Structural traps in the footwall of the thrust belts are developed and preserved well in terms of structural stability. The traps with rows and belts are developed by taking the gypsum-salt rocks in Jialingjiang Formation and Leikoupo Formation as caprock, the faults in thrust belt as the oil and gas migration passage (Fig. 5.11). 18 traps with an area of 370 km^2 are preliminarily determined. The reservoir evaluation and the implementation of the traps in the footwall of the faults should be strengthened to select the exploration targets.

The grain-beach karst reservoir in Leikoupo Formation in the southwest of Sichuan Basin is developed in large scale, and the effectiveness of trap is the key to successful exploration. The major gas reservoirs in Leikoupo Formationare mainly composed of carbonate grain beach (Plate III), which are mainly developed with two types of reservoir-permeability. The reservoir space of sand-clastic dolomite is mainly composed of remaining intergranular pores and intergranular dissolved pores, and the pore throat is mainly composed of necking throat and tubular throat, with good combination between pore and throat. And the reservoir is mostly characterized by medium porosity and medium permeability. The reservoir space of oolitic

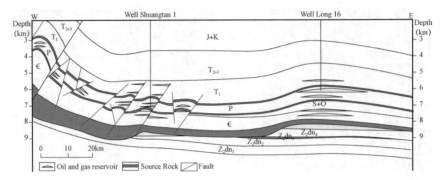

Fig. 5.11 Prediction profile of marine stratigraphic gas reservoir in northwest of Sichuan Basin

limestone and remaining bioclastic dolomite is mainly composed of intragranular dissolved pores, mould pores and organism pores, with low development of throat. The reservoir is mainly characterized by medium porosity and low permeability. The preservation of primary intergranular pores and the leaching of connate atmospheric fresh water are the key to the formation of these two types of grain beach reservoirs. According to logging evaluation, the reservoirs in Member 4 of Leikoupo Formation are developed with thickness of about 20–57 m, porosity of 3.72–4.8% and distribution area of 3500 km^2. The effectiveness of traps is the key to the success of exploration in Leikoupo Formation in the southwest of Sichuan Basin. Wells of Wu 1, Gaojia 1 and Lianhua 2 are lost in the hanging wall of thrust belt, but high yield is gained in Well Guankou 003–5 in the footwall. Significant discoveries in structures such as Yazihe and Xinchang are also gained in the structural stability area of thrust belt footwall. In a word, recent breakthroughs are all gained in the footwall structure of the thrust belt (Fig. 5.12). As a result, it is recommended to implement traps, select targets to drill in the structural stability area.

5.3   Geological Evaluation Technology of Deep Oil and Gas

The key problems of geological evaluation in deep oil and gas are the scale of deep oil and gas, the economy of resources, the parameter standard and method flow of favorable exploration zones evaluation and so on. The economic evaluation technology of deep oil and gas resources is formed by counting and establishing the oil and gas reservoir data related to geological conditions, surface engineering and market changes, determining parameter standards and method flow, and programming software module. On the basis above, the evaluation technology of favorable zones in deep exploration is formed by determining the evolution parameters, standards and method flow of deep favorable zones. And the scale and economy of deep oil and gas resources can be answered through this technology, which provides support for the potential evaluation of deep oil and gas resources, the evaluation of favorable zones and the implementation of exploration targets.

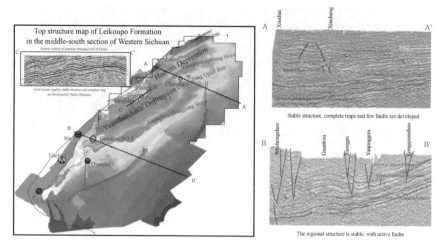

Fig. 5.12 Thrust structures and traps in southwest of Sichuan Basin

5.3.1 Economic Evaluation Technology of Deep Oil and Gas Resources

The key to the economic evaluation of oil and gas resources is to answer "Whether the evaluated resource is economical" "Whether development has economic value". Different from the evaluation of oil and gas resources, economic evaluation emphasizes whether there are advanced and applicable technologies in the exploration and development process in the predicted scale of oil and gas resources and the key is whether development can bring economic benefits. As a result, the strategic deployment of exploration can be provided by selecting resource areas with economic value by the economic evaluation of various types of resources.

Economic evaluation methods include cash flow and index evaluation. The former is mainly used for resource evaluation of mature exploration areas; while the latter can be used for all types of resource evaluation. In China, most of the deep oilfields are not in the middle-late stage of development according to the characteristics of economic evaluation of deep oil and gas resources. Based on the current actual situation in deep strata, the evaluation system of method selection and parameter standard in economic evaluation is established and the workflow and application effect of the evaluation are shown with cases, by selecting suitable economic evaluation of oil and gas resources through analysis on commonness and particularity of deep carbonate oil and gas resources economy.

(1) Evaluation Ideas

The index evaluation method, which is also known as the evaluation method of index element, is used for qualitative evaluation of the economic value of resource areas by statistics and expert evaluation methods based on the statistical analogy data.

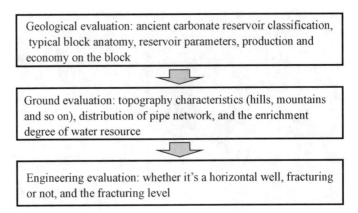

Fig. 5.13 Economic evaluation technology of deep ancient carbonate oil and gas resources

The evaluation method of economic index elements is put forward, which integrates geology, surface and engineering conditions, according to the characteristics of deep ancient carbonate rocks. On this basis, the evaluation process and index scheme are established for different reservoirs, such as karst carbonate reservoir, grain-beach carbonate reservoir and reef-beach carbonate reservoir (Fig. 5.13).

①   Evaluation Index

The evaluation index is an important factor to reflect the resource economy. The evaluation index is established under a unified logical framework according to various resources. In the project of evaluation index system, the evaluation of single well production should also be considered for some areas with relatively mature development in addition to geological factors, surface factors and technical factors. The index factors vary according to different types of resource.

According to the characteristics of deep oil and gas resources in China, the evaluation of deep clastic and volcanic reservoirs is not involved, as the oil and gas accumulation is mainly in carbonate reservoir. Index can be divided into single element index, or element for short, and multiple elements index, or index for short, consisting of several elements. Three types of index elements are developed, including digital-type, conceptual-type and fuzzy-type (Table 5.5). Among them, the basic data is mainly composed of digital-type, geological data is mainly composed of digital-type and conceptual-type, and ground data and technical data are mainly composed of conceptual-type and fuzzy-type.

Table 5.5 Basic data for economic evaluation of deep carbonate oil and gas resources

Basis	Evolution area	The central of Sichuan Basin	Maps
	Target layer	Dengying Formation, Longwangmiao Formation	
	Area (km^2)	8100	
	Resources (10^8 m^3)	15,000	
	Recoverable coefficient (%)	15	
	Resource abundance (10^8m^3/km^2)	3.3	
	Buried depth of target layer (m)	4000–5800	Buried depth map
	Recoverable resources (10^8 m^3)	10,000	Abundance map
Geology	Reservoir quality	(1) Marine type; (2) the thickness is 20–135 m, with stable horizontal distribution, and the average length in horizontal well of 1000 m; (3) the formation pressure coefficient is 1.1–2.0; (4) the porosity is 3.5–8%; (5) the permeability is 0.008–10 mD	
	Quality of caprock and trap	(1) Thickness of caprock (m); (2) distribution range of caprock; (3) types and area of traps	Trap morphology map
Ground	Topographic features	Mounds are mainly developed in Central Sichuan	Topographic map
	Pipeline, distance between resource area and pipeline (km)	Pipelines are laid in 50% of the evaluation area	Pipeline distribution
Technology	Drilling and completion technology of horizontal wells	The drilling and completion technologyof horizontal well are basically developed	
	Fracturing technology	Technology is basically developed	
Single Well	Initial daily production of single vertical well (m^3)	Production segmentation	Production of typical well
	Initial daily production of single horizontal well (m^3)		

Table 5.6 Classification standard table of index element evaluation

Grade of element evaluation	Good	Medium	Poor
Score	90–100	60–90	0–60
Price	High	Midium	Low
Oil price ($/bbl)	100–120	80–100	60–80
Gas price (¥/m³)	2–3	1–2	<1

② Index evaluation standard

The evaluation grade and reference standard in economic evaluation are given to different index elements of oil and gas reservoir, such as karst type, grain beach type and reef beach type, in order to give the economic meaning to index elements effectively. Among them, the evaluation of elements or indicators is divided into three grades: good (score greater than 90), medium (score 60–90) and poor (score less than 60); the economic analysis is divided into three grades according to the price (Table 5.6).

③ Economy evaluation of resource

The evaluation of resource economy is to estimate the resources economic value according to the predicted amount of resources by the method of expert evaluating and grading. The evaluation makes the following assumptions:

Q—resources in the evaluation area, 10^4 t or 10^8 m³;
Q_r—technical recoverable resources in evaluation area, 10^4 t or 10^8 m³;
Q_e—technical and economic resources in the evaluation area, 10^4 t or 10^8 m³;
E_i—the evaluation score of the i-th element, $i = 1.2 ..., n$;
n—number of elements;
R_i—weight coefficient of the ith element, decimal

(i) Preparation of basic data

First of all, prepare basic data and materials for the objects of economic evaluation. The content mainly includes the basic data of the evaluation area and the data required by the economic evaluation index standard. Among them, the basin name, evaluation area name and target strata are included in the basic data.

(ii) Index element evaluation

Each element of the evaluation area is graded with reference to the evaluation criteria of the index elements such as karst type, grain beach type and reef beach type according to the actual data of the evaluation area elements.

$$E_i = \begin{cases} 90 - 100, & the\,ith\,element \in Good \\ 60 - 90, & the\,ith\,element \in Medium \\ 0 - 60, & the\,ith\,element \in Poor \end{cases}$$

The economic grade of resource area is comprehensively evaluated by the economic evaluation of index element E_i. Three types of models for the quantification of index elements are developed, such as calculation method, qualitative description method and the combination method of qualitative description and graph (Xunan et al. 2010).

Calculation method.

In the index system, some element scores can be calculated from formula. This category mainly include the buried depth of the target layer, the recoverable resources and so on. For example, the quantitative evaluation method of buried depth of target layer is:

$$E_i = \begin{cases} 90 - 100, & h < 4500 \\ 60 - 90, & 4500 \le h < 5500 \\ 0 - 60, & 5500 \le h < 6500 \end{cases}$$

In the formula h—buried depth of target layer, m.

Qualitative description

Some index elements can not be quantified completely by formula, and should be determined by experience in the evaluation, mainly including uncertain index elements such as engineering technology. Taking the drilling and completion technology of horizontal well as an example. In order to determine the evaluation of the three levels, such as good, medium and poor, the specific scores should be given by the experienced experts to achieve the purpose of quantification.

Combination method of qualitative description and graph.

In order to intuitively analyze the economic value of resource areas, some index elements (such as reservoir quality) need to be quantified in combination with graphs. In the evaluation, such elements mainly include index elements such as terrain and geomorphology of the ground conditions, and the score needs to be determined with the help of graphs.

(3). Economic resource evaluation

Resource economy evaluation is affected by geology, surface, technology, single well and other indicators, and each indicator is affected by multiple elements. In the evaluation, the resource economy is evaluated on the basis of elements.

$$Q_e = Q \times \left[\sum (E_i \times R_i) \right] \times \text{ratio of resource economy conversion coefficient}$$

In the formula, the ratio of resource economy conversion coefficient is estimated according to statistics.

(2)   Technical workflow of evaluation and standard of parameter

The evaluation of index elements is mainly carried out in the economic evaluation, which is combined with the characteristics of resource evaluation based on the method and technology of economic evaluation.

The key to carry out the evaluation of economic index elements is to understand the basic data, standardize the evaluation workflow and compare the evaluation results. The economic evaluation is carried out by the workflow as shown in Fig. 5.14, after determining the evaluation unit and major strata for a certain type of oil and gas resources. The main contents are as follows:

①   Preparation of basic data. Data preparation varies with different methods. The indicators such as geological conditions, surface conditions, engineering technology and single well production, as well as data and supporting maps of related elements are needed for the index evaluation method.
②   Economic evaluation. Economic evaluation results can be predicted by economic index evaluation or cash flow evaluation method. Among them, the conclusion of economic index evaluation is the score of economic evaluation, which is evaluated according to three grades: good (90–100), medium (60–90) and poor (0–60).
③   Comparative analysis of economic evaluation conclusions. The effective economic zone, economic zone and invalid economic zone of various resources are comprehensively evaluated based on the summary and comparison of economic evaluation results of different resources.

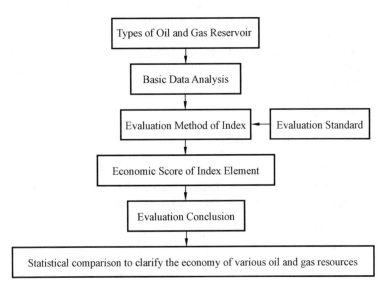

Fig. 5.14 Workflow of economic evaluation

5.3.2 Evaluation Examples

According to the types of oil and gas reservoirs of karst, grain beach and reef beach, the economic evaluation of deep ancient carbonate reservoirs in Gaoshiti-Moxi area in Sichuan Basin is carried out. Among them, the evaluation of karst reservoirs in Sichuan Basin is divided into six evaluation areas, including Gaoshiti structural area, Moxi structural area, Longnvsi structural area, East Gaoshiti structural area, Hebaochang structural area, inner depression structural area of Mianzhu-Changning (Table 5.7 and Fig. 5.15).

The economic index evaluation is carried out in the evaluation area by the preparation of basic data and maps above mentioned, the selection of evaluation methods, the establishment of evaluation index, the evaluation of economy, the analysis of conclusion and so on.

The first step is to prepare basic data. The basic data of the evaluation units are prepared by the economic evaluation indicators and elements. The basic maps are prepared mainly according to three evaluation areas, including Gaoshiti area, Moxi area and Longnvsi area (Table 5.8). Taking Longnvsi evaluation area as an example, the structural area is 752.31 km^2.

The second step is the interface and operation of economic evaluation system. The interface operation is realized by programming software.

The third step is the evaluation conclusion. The economic evaluation is carried out for different structural units in the paleouplift of deep basin in the central of Sichuan Basin by the proposed index evaluation of economic carbonate reservoirs in deep strata. The target layers include Dengying Formation and Longwangmiao Formation. The evaluation results are shown in Fig. 5.15.

5.3.3 Evaluation Technology of Favorable Zones for Deep Oil and Gas Exploration

The main evaluation purpose of oil and gas enrichment zone is to reduce the risk and improve the efficiency with the scientifical guidance of strategy and plan of oil and gas exploration. The evaluation method and result should be a process of dynamic change and continuous improvement because with the improvement of exploration degree and the increase of geological data, as well as the progress of various oil and gas exploration technologies and the deepening of geological understanding. The previous exploration results and cognition degree should be modified and improved in multiple rounds to realize the integration of zone evaluation with basic geological research and exploration practices, and ensure the rationality and scientificity of the results of oil and gas zone evaluation. As a result, the potential of oil and gas resources is tapped to the maximum extent for the effective selection of oil and gas exploration zones. Generally speaking, the evaluation system of oil and gas enrichment zone is developed with three parts: geological evaluation, resource prediction and economic

Table 5.7 Basic structural parameters of economic evaluation units of karst reservoirs in Sichuan Basin

Region	Area (km^2)	Structural location	Typical wells	Development characteristics of reservoir
Scale area 1	2041.90	Gaoshiti structure	Gaoshi 1, Gaoshi 2, Gaoshi 3, Gaoshi 6, Gaoshi 7, Gaoshi 8, Gaoshi 9, Gaoshi 10, Gaoshi 11, Gaoshi 12, Gaoshi 18, Gaoke 1	Two sets of reservoirs in Member 2 and 4 of Dengying Formation are developed, with mounds and beaches in platform margin. In the epigenetic stage it lies in karst slope. The fractures promote the karst
Scale area 2	1757.40	Moxi structure	Moxi 8, Moxi 9, Moxi 10, Moxi 11, Moxi 12, Moxi 13, Moxi 17, Moxi 18, Moxi 19, Moxi 21, Moxi 22, Moxi 47, Anping 1	Two sets of reservoirs in Member 2 and 4 of Dengying Formation are developed, mainly composed of mounds and beaches in platform margin. The grain beach and platform flat are gradually developed in the eastern part. In the epigenetic stage it is formed in karst slope with development of fractures
Scale area 3	1518.41	Longnvsi structure	Moxi 23, Moxi 41	Two sets of reservoirs in Member 2 and 4 of Dengying Formation are developed, mainly composed of grain beach and platform flat. The eastern area is located in the karst highland, with relatively developed faults

(continued)

evaluation. The exploration practice and large number of researches on the deep marine carbonate reservoirs onshore show that, the distribution of large-scale oil and gas fields is controlled by paleo rift, paleo uplift, paleo slope, paleo platform margin and paleo fault zone. The following focuses on the methods and techniques of geological evaluation of carbonate oil and gas zones in three major basins.

Table 5.7 (continued)

Region	Area (km²)	Structural location	Typical wells	Development characteristics of reservoir
Scale area 4	752.31	East Gaoshiti structure	Gaoshi 21, Gaoshi 105	Two sets of reservoirs in Member 2 and 4 of Dengying Formation are developed, mainly composed of platform flat and lagoon. Grain beach is locally developed, and main body is developed in karst highland
Scale area 5	459.69	Hebaochang structure, Panlongchang structure	Pan 1, Heshen 1	Two sets of reservoirs in Member 2 and 4 of Dengying Formation are developed, with mounds and beaches in platform margin. The eastern part of the evaluation area is located in the karst highland, and the western part is located in the karst slope
Scale area 6	978.49	Inner craton rift of Mianzhu-Changning	Gaoshi 17	The reservoirs are not developed on the whole, and only developed in Member 2 of Dengying Formation in the area near Well Heshen 1. The reservoirs are developed mainly with mound beach in the platform margin, and in epigenetic stage it is developed in karst slope

(1) Evaluation principle

The principle of oil–gas zone division is to control the similarity of key geological elements and the predictability of evaluation results of oil and gas zone, by ensuring the integrity of regional division and classification as well as the similarity of geological conditions and characteristics of oil and gas accumulation, through the genetic mechanism of oil and gas zone based on the division of regional primary

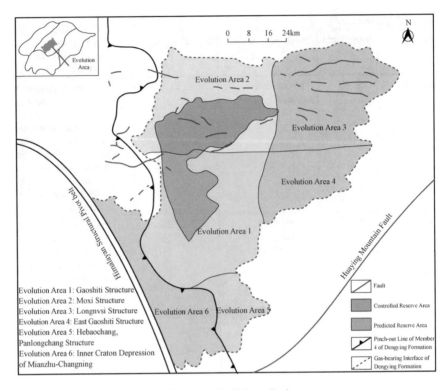

Fig. 5.15 Evaluation conclusion of the central in Sichuan Basin

and secondary structural units. The integrity of regional division and its classification refers to that they should include all types of oil and gas reservoirs in
carbonate field.The geological conditions and characteristics of oil and gas accumulation are basically similar, mainly referring to the similarity of sedimentary background, hydrocarbon source conditions, reservoir characteristics, oil and gas transportation system, caprock and preservation conditions and accumulation process,
but not emphasizing the similarity of oil and gas sources. The similarity of the key
geological factors in the oil and gas zone mainly refers to the similarity of the key
geological factors controlling the oil and gas accumulation in the same zone, with
certainty and comparability, such as paleouplift, reef beach in platform margin, overburden unconformity, paleo buried hill and fault zone. The results of regional division
and evaluation should be consistent with the existing exploration results and have a
predictive effect.

(2)   Evaluation method and workflow

The process of regional division and geological evaluation in carbonate oil and gas
zones is complex. The comprehensive method combined of vertical stratifications of
"reservoir combination" and plane zones of "key elements of oil and gas zone" is
taken with the following eight steps:

Table 5.8 Basic data for economic evaluation of carbonate oil and gas resources in deep strata

Basis	Area name	Longnvsi area of the central of Sichuan Basin	Maps
	Target layer	Dengying Formation	
	Area (km^2)	752.31	
	Resources (10^8 m^3)	2000	
	Recoverable coefficient (%)	15	
	Resource abundance (10^8 m^3/km^2)	2.8	
Geology	Buried depth of target layer (m)	5000–6500	Buried depth map
	Recoverable resources (10^8 m^3)	253	Abundance map
	Reservoir quality	(1) Marine; (2) the thickness is 20–135 m, with stable horizontal distribution, and the average length in horizontal well of 1000 m; (3) the formation pressure coefficient is 1.1–2.0; (4) the porosity is 3.5–8%; (5) the permeability is 0.008–10 mD	
	Quality of caprock and trap	(1) Thickness of caprock (m); (2) distribution of caprock; (3) types and area of traps	Trap morphology map
Ground	Topographic features	Small hills are mainly developed in the Central of Sichuan Basin	Topographic map
	Pipeline, distance between resource area and pipeline (km)	Pipelines are laid in 80% of the evolution area, while not laid in 20% of the evolution area	Pipeline distribution map
Technology	Drilling and completion technology of horizontal wells	The drilling and completion technology of horizontal well is basically developed	
	Fracturing technology	Technology is basically developed	
Single well	Initial daily production of single vertical well (m^3)	Production segmentation	Production of typical well
	Initial daily production of single horizontal well (m^3)		

(1) Vertical division and evaluation of reservoir combination;
(2) Analysis of key geological controlling factors in oil and gas zones;
(3) Plane division and evaluation of oil and gas zones;
(4) Estimation of the area and the scale of resource reserves of oil and gas zones in different reservoir combinations;
(5) Optimization of evaluation methods for oil and gas zones;

(6)   Establishment of evaluation parameter system and quantitative standard of oil and gas zones;

(7)   According to the quantitative standard of evaluation parameters in zones, the comprehensive evaluation values of the zones are obtained by comprehensive grading or "multiple map superposition" with the assignment of corresponding evaluation scores and weighted superposition in oil and gas zones of different reservoir combinations.

(8)   The queuing optimization and grading evaluation are carried out by comprehensive evaluation value combined with the exploration degree and resource potential of oil and gas zone.

(3)   Standards and Types of Regional Division

The evaluation standard of oil and gas zones in the study area should be optimized and established by the combination of overall exploration status, the spatial distribution of resources and exploration economic benefits, through the establishment of geological evaluation standard by key geological parameters of oil and gas accumulation in the scale area, which is firstly selected as the geological parameters of oil and gas zone evaluation in deep carbonate. It should be emphasized that the establishment of evaluation standard should focus on the main geological factors of accumulation, engineering and technical difficulties, exploration benefits and so on. Evaluation parameter standards and weights in different carbonate oil and gas enrichment zones are different. In general, common parameters and standards for evaluation of carbonate oil and gas zones are shown in Table 5.9.

The boundaries of deep carbonate oil and gas zones take "oil–gas accumulation combination" as the basic geological unit vertically, and the boundaries on the plane can be developed as: structural unit boundary of basin, abrupt zone of structure, boundary of variational reservoir lithology or physical property, pinch-out belt of fault or stratum and unit boundary of oil–gas migration and accumulation.

As the basic geological unit of vertical stratification evaluation of carbonate oil and gas zones, the geological conditions and characteristics in the same accumulation combination are basically similar, such as the sedimentary background, source conditions, reservoir types and characteristics, oil and gas passage system, caprock and preservation conditions and accumulation process. The accumulation combinations can be classified and named according to two key elements of source-reservoir combination relationship and reservoir type, with no emphasis on the unity of oil and gas sources.

At present, five kinds of spatial relations of source-reservoir combination in carbonate oil and gas accumulation have been found, such as type of self-generation and self-storage, near-source type of lower-generation and upper storage, far-source type of lower-generation and upper-storage, type of para-generation and lateral-storage, as well as type of upper-generation and lower-storage. Oil and gas exploration shows that carbonate reservoirs are mainly divided into three genetic types, such as reef beach reservoir, dolomite reservoir and karst reservoir. According to the five source-reservoir combination and three major genetic types of reservoirs, 13

Table 5.9 Major parameters and standards for geological evaluation of deep carbonate oil and gas zones

Evaluation parameters					Weight		
key parameters	Weight	subclass	Weight		I	II	III
					1–0.7	0.7–0.4	<0.4
Source conditions	0.15	Types of source rocks	0.3		Black mudstone and shale	Grey mudstone	Grey muddy dolomite
		Layers of source rocks	0.05		More than two sets	Two sets	One set
		Thickness of source rocks (m)	0.2		≥ 500	500–50	<50
		Hydrocarbon-generating intensity (10^4 t /km^2)	0.2		>1000	1000–200	<200
		Distance from hydrocarbon generation kitchen(km)	0.25		<10	10–50	>50
Passage system	0.1	Fault length(km)	0.2		3	3–1	<1
		Fracture density(strip/m)	0.2		0.5	0.5–0.05	<0.05
		Fracture properties and opening	0.2		Diagonal tension joint of high angle	Unfilled vertical tension joint	Shear joint
		Unconformity and composite transportation	0.2		Good connectivity	Preferable connectivity	Poor
		Matching of transport elements	0.2		Good matching	Preferable matching	Poor matching
Reservoir conditions	0.15	Sedimentary facies	0.3		Reef breach in platform margin	Beach in platform	Platform
		Thickness of reservoirs(m)	0.4		>50	10–50	<10

(continued)

Table 5.9 (continued)

Evaluation parameters						
key parameters	Weight	subclass	Weight	I	II	III
				1–0.7	0.7–0.4	<0.4
Tectonic setting	0.05	Types of reservoirs	0.3	Fracture-cave dolomite/limestone	Pore-cave dolomite/limestone	Pore-fracture limestone
		Location of paleostructure	0.6	Stable paleouplift and slope	Active paleouplift and upper slope	Lower slope and depression
		Structure location at present	0.4	Anticlinal zone	Pivot belt	Syncline area
Trap conditions	0.1	Types of traps	0.3	Structural type	Composition of structure, lithology and strata	Lithology-stratigraphy
		Coefficient of trap area (%)	0.3	>30	10–30	<10
		Trap amplitude (m)	0.4	>400	50–400	<50
Source-reservoir combination	0.15	Types of source -reservoir combination	0.6	Type of self-generation and self-storage, near-source type of lower-generation and upper-storage, type of para-generation and lateral-storage, type of upper-generation and lower-storage	Far-source type of lower-generation and upper-storage, type of upper-generation and lower-storage	Far-source type of lower-generation and upper-storage, type of upper-generation and lower-storage

(continued)

Table 5.9 (continued)

Evaluation parameters			Weight	I	II	III
key parameters	Weight	subclass	Weight	1–0.7	0.7–0.4	<0.4
Preservation conditions	0.05	Spatial contact relationship	0.4	Surface contact of direct superimposition of upper and lower part	Surface or line contact of upper and lower migration	The upper and lower layers of the partition are overlapped or moved
		Lithology of caprock	0.3	Development of thick mudstone of gypsum-salt rock	Medium thick mudstone	Carbonate rock
		Thickness of caprock (m)	0.3	≥500	100–500	<100
		Damage degree	0.4	No damage	Minor damage	More serious damage
Time and space allocation of accumulation factors	0.1		1.0	Early or simultaneous	At the same time	Late
Reserve scale of regional resources	0.05	Regional resources(10^4 t)	0.4	>10,000	2000–10,000	<2000
	0.1	Zone area(km^2)	0.6	>300	100–300	<100

Table 5.10 Common types of accumulation combinationin of carbonate strata

Types of reservoir relationship of source-reservoir combination	Reef breach	Dolomitic	Fracture, cave and karst
Self-generation and self-storage	Reef-breach reservoir of self-generation and self-storage		
Near-source of lower-generation and upper-storage	Reef-breach reservoir of lower-generation, upper-storage and near-source	Dolomitic reservoir of lower-generation, upper-storage and near-source	Karst reservoir of lower-generation, upper-storage and near-source
Far-source of lower-generation and upper-storage	Reef-breach reservoir of far-source of lower-generation, upper-storage	Dolomitic reservoir of far-source of lower-generation, upper-storage	Karst reservoir of far-source of lower-generation, upper-storage
Para-generation and lateral-storage	Reef-breach reservoir of para-generation and lateral-storage	Dolomitic reservoir of para-generation and lateral-storage	Karst reservoir of para-generation and lateral-storage
Upper-generation and lower-storage	Reef-breach reservoir of upper-generation and lower-storage	Dolomitic reservoir of upper-generation and lower-storage	Karst reservoir of upper-generation and lower-storage

common types of accumulation combination can be obtained through the combination of the source-reservoir combination and genetic types, as shown in Table 5.10.

Affected by structural formation mechanism and structural unit division of different basin types (craton basin, depression basin and fault basin), carbonate oil and gas zones can be divided into four major types, such as structure-controlled zone, stratigraphy-controlled zone, lithology-controlled zone and compound -controlled zone. Each zone can be divided into subzones (Table 5.11).

5.3.4　Evaluation of Favorable Exploration Zones in Typical Areas

Taking the reservoirs in Sinian-Lower Paleozoic and Qixia-Maokou Formation of Lower Permian in Sichuan Basin as key area, the rapid development of deep oil and gas exploration can be promoted by deepening the geological understanding.

(1)　Exploration Direction and Favorable Zone of Dengying Formation of Sinian

The major accumulation combinations in Dengying Formation of Sinian in Sichuan Basin are mainly karst reservoir of para-generation and lateral-storage, sandwich-type, near-source dolomite reservoir of upper-generation and lower-storage, as well as karst reservoir of lower-generation and upper-storage. The major source rocks are

Table 5.11 Types of deep carbonate oil and gas zones

Types of Basin	Primary structural units	Second structural units	Types of oil and gas zones			
			Structure-controlled zone	Stratigraphy-controlled zone	Lithology-controlled zone	Compound-controlled zone
Craton carbonate platform	Uplift rift trough, depression, slope Depression steep hill	Uplift, depression, gentle slope and thrust belt	Anticline zone, fault/fracture zone, structural pivot belt, rock perforation zone	Paleo buried hill belt, denudation unconformity belt, overburden unconformity belt, platform-margin reef/beach belt, inner platform reef/beach belt	Lithologic mutation zone, slope zone and dissolution transformation zone	Karst- buried hill zone, lithology-anticline zone, stratum-fracture zone, stratum-anticline zone, lithology-structure pivot belt

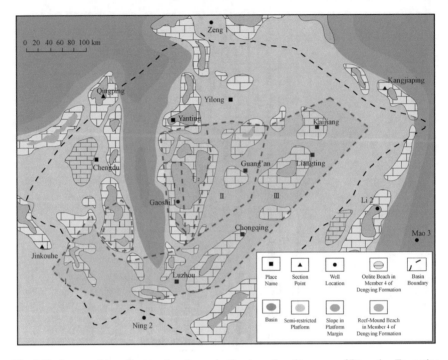

Fig. 5.16 Superposition of comprehensive evaluation in exploration zones of Dengying Formation of Sinian, Sichuan Basin

developed in Qiongzhusi Formation of Cambrian, while the reservoirs are developed in Member 2 and 4 of Dengying Formation. Three major favorable exploration zones are divided in Dengying Formation of Sinian in Sichuan Basin by the distribution characteristics of hydrocarbon source rocks, the development scale of mound-shoal bodies, and the matching relationship between source rocks and reservoirs, based on the analysis of reservoir forming control of Deyang-Anyue rifting.

According to the advantages and disadvantages of accumulation factors, the exploration zones (Fig. 5.16) can be divided into: Zone I (surrounding "rift trough" structural pivot belt), Zone II (slope belt), Zone III (structural belt of high-steep faulted anticline), of which Zone I is the most favorable exploration area at present.

5.3.5   Zone I (Surrounding "Rift Trough" Structural Pivot Belt)

This exploration zone surrounds Deyang-Anyue rift trough (Fig. 5.16), with the area of about 1.6×10^4 km^2. At present, gas reservoirs of Sinian and Weiyuan Ziyang, Member 2 of Dengying Formation in Gaoshiti-Moxi area, and Memer 4 of Dengying Formation in Gaoshiti-Moxi-Longnvsi area have been found. Among them, 25 wells have been drilled in Dengying Formation of Gaoshiti-Moxi area, as well as 17 wells gas testing and obtaining industrial gas. This exploration zone can be divided into three favorable blocks according to the difference of exploration degree and accumulation conditions. Zone I_1 is located in the platform eastern margin of the "rift trough", where the karst reservoir of mound-beach facies is developed with large thickness; the strata of Lower Cambrian in the "rift trough" are developed with huge-thick high-quality source rocks, which are laterally connected with high-quality reservoirs of Sinian and favorable for oil and gas accumulation;it is located in the high part of the inherited paleouplift, with the good superposition of paleostructure and present structure, and always in the direction area of hydrocarbon migration; The gas reservoirs developed in Member 2 and 4 of Dengying Formation in Gaoshiti-Moxi-Longnvsi area have been found, with good exploration effect and high-yield gas wells. Zone I_2 is located in the eastern of platform margin of Gaoshiti-Moxi area, with an area of about 6000 km^2. It is now located in the slope developed from paleouplift. Gas is obtained through the oil testing of Member 4 of Dengying Formation in Well Nvji and Moxi 23, which shows the low exploration degree and great exploration potential. Zone I_3 is located in platform margin of western of "rift trough" in Deyang-Anyue Depression, which the present is high structural position developed from western paleouplift. The strata of Member 4 of Dengying Formation are strongly eroded, with thin residual thickness and local formation lack. The Member 2 of Dengying Formation is the major production layer. A large amount of natural gas lost by the damage of caprock, due to the large extent of structural uplift caused by the effect of Himalayan period.

② Zone II (slope zone)

Zone II non is located in the structural slope area developed from paleouplift to the east of Gaoshiti-Moxi structure. It is east to the western of Guang'an, north to the northern of Nanchong and south to Hechuan, with an area of about 1.3×10^4 km^2. At present, three wells have been drilled in this area, which reflect the favorable accumulation conditions as: (1) during the evolution of paleouplift, the structural deformation in this area is weak, and structural traps such as Nanchong, Guang'an and Longnvsi are developed. (2) Mounds and beaches are developed in Dengying Formation. The whole research area is developed in the karst slope belt of top Sinian before deposition in Cambrian. The mounds and beaches are transformed by regional karstification, and the reservoirs are generally developed. (3) The reservoirs are directly contacted with the high-quality overlying source rocks of Cambrian, with the development of source-reservoir combinations, such as lateral connection, upper-generation and lower-storage. Regional unconformity provides favorable passage for oil and gas migration and accumulation. (4) Gas is generally found in the Member 4 of Dengying Formation in Gaoshiti-Moxi area with shallow buried depth.

③   Zone III (structural belt of high-steep faulted anticline)

This area is located in the structural belt of high-steep thrust short anticlinein in the south and east Sichuan area around the paleouplift of central Sichuan Basin, with a total area of about 4.5×10^4 km^2. 6 wells have been drilled in Ziliujing, Dawoding, Laolongba, Tiangongtang, Hanwangchang and Panlongchang of the south Sichuan Basin, with favorable accumulation conditions, mainly reflected in: (1) the development of karst reservoirs of mound beach in the platform of Dengying Formation of Sinian; (2) large amount of structural traps of Himalayan; (3) the exploration of Well Heshen 1 proves that the traps in Himalayan can capture late-stage cracking gas (kerogen cracking and dispersed liquid hydrocarbon cracking) to form gas reservoirs.

(2)   Exploration direction and Favorable Zones of Longwangmiao Formation in Cambrian

The reservoir combination of Longwangmiao Formation of Cambrian is mainly near-source dolomite reservoirs of lower-generation and upper storage, with the major gas sources from Qiongzhusi Formation of Cambrian. Based on the current exploration results and the major understanding of controlling factors in gas accumulation (reservoir and unconformity, fault passage system), the Longwangmiao Formation of Cambrian can be divided into three major exploration zones by the distribution of hydrocarbon source kitchen of Qiongzhusi Formation, structural background and spatial distribution of reservoir to divide the accumulation combinations, and the boundary of migration and accumulation unit. According to the advantages and disadvantages of the accumulation factors, the zones can be divided as: Zone I (structural pivot belt), Zone II (slope belt), and Zone III (high-steep structural belt). Among them, Zone I is the most favorable exploration area at present.

①   Zone I (structural pivot belt arounding "rift trough")

The exploration area is located on the top of the high-quality hydrocarbon source kitchen of Deyang-Anyue Depression, where is the high part of the inherited paleouplift with stable structure and reservoirs of grain beach, which is the superposition of ancient and modern structures and is conducive to the formation and preservation of the primary gas reservoirs. As the most favorable exploration area for the gas exploration in Longwangmiao Formation of Cambrian, the exploration area is developed with the area of about 8000 km^2 (Fig. 5.17). At present, three gas reservoirs of Longwangmiao Formation have been found, suh as Gaoshiti, Moxi and Longnvsi, with 21 industrial gas wells of 33 oil testing wells. The exploration area can be divided into three favorable blocks based on the difference in exploration degree and accumulation conditions. Zone I$_1$ is located in the high part of the paleouplift of "rift trough", mainly in Moxi-Longnvsi structure, with an area of about 2400 km^2 and favorable accumulation conditions, which are reflected in: (1) it lies in the superimposed area of denuded paleouplift and karst reservoir of thick grainbeach facies; (2) it is near the hydrocarbon source, and located above the high-quality

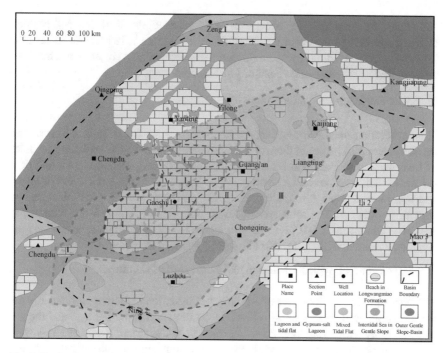

Fig. 5.17 Comprehensive evaluation superposition of exploration zone of Longwangmiao Formation of Cambrian in Sichuan Basin

source kitchen of Qiongzhusi Formation in Deyang-Anyue Depression; (3) fractures in hydrocarbon source are developed with superior hydrocarbon supply conditions; (4) it lies in ancient and modern structural superposition area, where it is the target area of long-term hydrocarbon migration; (5) drilling results are mainly high-yield wells, confirming the large scale of oil and gas accumulation, and the proved reserves in this area are 4403.83×10^8 m^3. Zone I$_2$ is located in the Gaoshiti area on the wing of the paleouplift, with an area of about 2000 km^2. The accumulation conditions are similar to Zone I$_1$, but relatively thin reservoir thickness and relatively good exploration effect. Zone I$_3$ is located in the northern Moxi-Longnvsi structure, where is north to Nanchong area, with an area of 3600 km^2. It is located in the high part of paleouplift slope, and develops mainly lithologic gas reservoirs. Good shows have been encountered in Longwangmiao Formation in Well Nanchong 1.

② Zone II (exploration area of slope belt)

The exploration area is located in the upper slope of paleouplift in the east and south of Zone I. At present, 3 exploration wells are drilled in Guang'an, Hebaochang and Panlongchang structures, and producing water. Wells in Longwangmiao Formation in Weiyuan are drilled with gas-bearing strata, with exploration area of 1.9×10^4 km^2. The favorable reservoir conditions are shown as follows: (1) the development of grain beach facies; (2) the development of traps, such as Weiyuan, Hebaochang, Nanchong and Guang'an structures; (3) except for Weiyuan area, the structural deformation of the slope belt in the eastern paleouplift is weak, and it is possible to develop lithologic, structural-lithologic and structural gas reservoirs. At present, gas has been encountered in Weiyuan and Nanchong structures.

③   Zone III (high-steep structural belt)

The exploration area is located in the high-steep structural belts of southern and eastern paleouplift in the southern and eastern Sichuan Basin. The sedimentary facies of Longwangmiao Formation is mainly composed of evaporation lagoon and tidal flat facies. The gypsum-salt reservoirs are developed with large thickness of layers and thin of reservoirs. It is mainly developed as lithologic gas reservoir with low exploration degree. The total area is about 6.2×10^4 km^2 and the natural gas resources can reach $(0.5–0.7) \times 10^8$ m^3, which has certain exploration risk and the key factor is whether the reservoir has scale development.

(3)   Exploration Prospect of Xixiangchi Formation of Cambrian

The accumulation combination of Xixiangchi Formation of Cambrian is mainly the far-source dolomite reservoir with lower-generation and upper-storage, and the gas source mainly come from Qiongzhusi Formation of Cambrian. Based on the current exploration results and the main understanding of controlling factors of gas reservoir, the Xixiangchi Formation of Cambrian can be divided into three major exploration zones (Fig. 5.18) by the main controlling factors of gas accumulation, such as passage system, gas source supply and grain beach dolomite, and combined with structural background structural background the accumulation combination to divided and the boundary of migration-accumulation units. According to the advantages and disadvantages of accumulation elements, the areas can be divided as: Zone I (structural pivot belt), Zone II (slope zone), Zone III (high-steep structural zone).

Zone I is mainly located in the high part of the paleouplift, with an area of about 3.3×10^4 km^2. The superior accumulation conditions are mainly reflected in the development of beach reservoirs and present structural traps, such as 2.11×10^4 m^3 of daily gas in the Xixiangchi Formation of Well Moxi 23, 13.64×10^4 m^3 of daily gas in the Xixiangchi Formation of Well Nanchong 1 in Nanchong structure. Zone II is located in the south slope of the paleouplift, with an exploration area of about 1.8×10^4 km^2, and the development of lithologic traps. Zone III is located in the high-steep structural belt in eastern and southern Sichuan Basin, with an area of about 3.9×10^4 km^2. The high-steep structure-fault anticline traps are developed above the gypsum-salt reservoirs in Gaotai Formation, formed as accumulation combination of paragneration and lateral-storage with Longmaxi Formation of Silurian, and developed as fault and fracture structures related to the gypsum-salt rocks. The accumulation developed in the late stage, and with no exploration well at present, which needs to be confirmed by drilling.

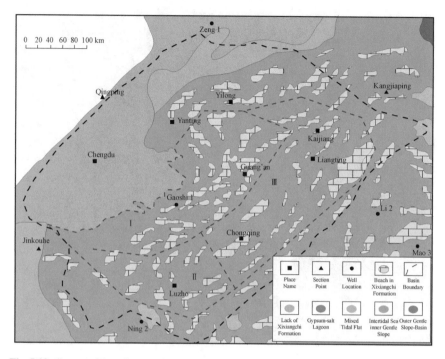

Fig. 5.18 Superposition of comprehensive evaluation of exploration zone in Xixiangchi Formation of Cambrian, Sichuan Basin

(4)   Evaluation of the Favorable Zone of Qixia-Maokou Formation

The natural gas resources of the Middle Permian in Sichuan Basin can reach 1.47×10^{12} m^3, but the overall exploration degree is very low. However, a good exploration prospect of the Middle Permian in the basin is shown by the high yield in the porous dolomite reservoirs of Well Shuangtan 1, Moxi 31X1, Nanchong 1 and Moxi 39. The research on the Maokou Formation shows that, karst fracture-cave reservoirs can be developed in the karst highland and the slope area, with large exploration area in the basin. The large-scale dolomite reservoirs in Qixia Formation and Maokou Formation, as well as the potential for large-scale oil and gas accumulation can be confirmed by the exploration breakthrough in dolomite reservoirs of beach facies. It shows that the most favorable area for gas exploration in the Middle Permian is the superposition area of karst reservoir and dolomite reservoir of beach facies based on the research ideas of overall exploration and large-scale exploration.

The large-scale karst reservoirs in Middle Permian are mainly effected by the tectonic movements of Dongwu Period of late Middle Triassic in Sichuan Basin, and the development of karst landform in the Middle Triassic is controlled by the development of large-scale Luzhou-Tongjiang paleouplift (Fig. 5.19). The reservoirs in Maokou Formation of Sichuan Basin can be divided into four favorable exploration zones by preservation conditions, structure characteristics and combination of source,

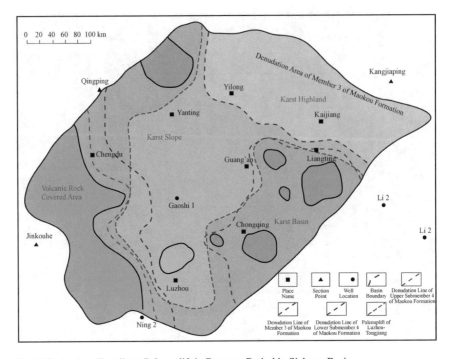

Fig. 5.19 Luzhou-Tongjiang Paleouplift in Dongwu Period in Sichuan Basin

reservoir and cap, such as the northwest Sichuan Basin, southwest Sichuan Basin, central-south Sichuan Basin and northeast Sichuan Basin. All the areas are located in the karst slope or karst highland, and the karst reservoirs should be relatively developed, with hydrocarbon generation intensity of more than 20×10^8 m^3/km^2 and relatively great exploration potential. Among them, hydrocarbon can be supplied laterally by two sets of source rocks in the favorable exploration area of the central-south Sichuan Basin, which are located in the karst highland. The oil and gas fields of large and middle scale can be formed by good preservation with low amplitude but large scale of structural traps. The second favorable exploration area lies in the northeast Sichuan Basin, with high gas generation intensity of Silurian source rocks, the development of structures, the high amplitude of traps and the development of karst reservoir, but worse preservation condition than that of the central Sichuan Basin. The last favorable exploration area is located in the northwest and southwest Sichuan Basin, which supplies hydrocarbon from the source rock of Middle Permian. The structures are developed in karst highland and slope area, which are conducive to the development of karst reservoir. However, it is necessary to optimize the structure of mountain belt for drilling due to the poor preservation condition caused by the location before Longmen Mountain and Micang Mountain.

(5) Favorable Exploration Targets of combination of Gypsum-Salt Rocks and Carbonate Rocks in East Sichuan Basin

Many important discoveries have been made in the global under-salt exploration. Three sets of source rocks, three sets of reservoirs, two sets of regional caprocks and three sets of source-reservoir-cap combination have been developed in the deep under-salt strata of Sinian-Cambrian in East Sichuan Basin. At the same time, the Xuanhan-Kaijiang paleouplift has also been developed, with favorable natural gas accumulation conditions. The high-quality reservoirs are developed in upper-salt strata of Xixiangchi Formation of Cambrian, and high-quality source rocks are developed in Silurian. The accumulation model with good lateral connection of source rocks and reservoirs can be formed by the communication of faults. 6–7 rows of large structural traps in the deep strata of East Sichuan Basin are developed from west to east, which have favorable conditions for the formation of large-scale gas reservoir, and developed as an important strategic succeeding exploration field in future.

Based on the analysis of nature gas accumulation conditions and source-reservoir-cap combination, the mud shale of Shuijingtuo Formation is the most important source rock of deep under-salt strata in the east Sichuan Basin. The weathering crust and interlayer karst of Member 4 of Dengying Formation, as well as the grain beach and weathering crust karst of Shilongdong Formation are important reservoirs, which all closely related to the source rocks. However, part of the reservoirs in Shilongdong Formation, especially reservoirs of grain beach, are developed less scale than that of Member 4 of Dengying Formation, such as the gypsum-salt layers developed near Well Zuo 3. As a result, the first choice for exploration in deep under-salt strata of East Sichuan Basin should be reservoirs in Member 4 of Dengying Formation and beach bodies of Shilongdong Formation of paleouplift and slope, and followed by reservoirs in Member 2 of Dengying Formation. The source rocks of Silurian are well developed in upper-salt strata in East Sichuan Basin, and they can form the laterally connected accumulation model with reservoirs in Xixiangchi Formation of Cambrian. As a result, the Xixiangchi Formation can be the major exploration strata in upper-salt area in East Sichuan Basin.

According to the major direction of oil and gas migration, the East Sichuan Basin has always been located in the slope area between the paleouplift of Leshan-Longnvsi, and hydrocarbon generation depression of Chengkou in Northeast Sichuan and hydrocarbon generation depression in West Hunan and Hubei in the geological history, that is, the long-stage favorable direction area of oil and gas migration (Fig. 5.20). The most favorable oil and gas accumulation area in the East Sichuan Basin is to the west of Well Zuo 3-Liangping-Kaijiang area, which is closely adjacent to central Sichuan Basin. The major gas supply in deep under-salt strata of East Sichuan Basin are two types, such as crude oil into gas in paleoreservoir cracking, and soluble organic matters detained and dispersed in the source rocks cracking during the high-over-mature stage.

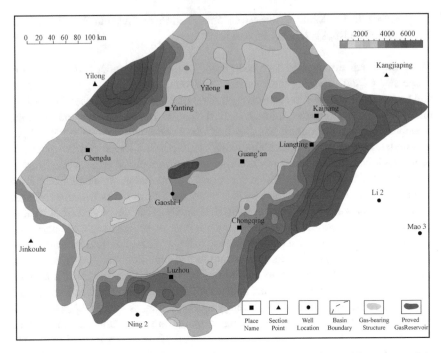

Fig. 5.20 Paleostructural map of the top interface and superposition of favorable gas reservoir in Sinian before Jurassic in Sichuan Basin

①   Three Favorable Zones in Under-Salt Area of the East Sichuan Basin

Paleouplift and slope in Xuanhan-Kaijiang area.

In this area, the grain beach and karst weathering crust reservoirs are widely developed, and the gypsum-salt caprocks in Cambrian are distributed in large scale. The structural traps are distributed in large scale and in rows, but far away from the center of source rocks.

Thrust-fold belt in the front of Daba Mountain.

This area is adjacent to the center of hydrocarbon source, and located in the slope area of paleouplift, close to the platform margin, and developed with beach body. It is difficult to identify the traps because of the poor quality of seismic data and the small area. And the preservation condition is poor because of the complex structures.

Thrust-fold belt in the front of Qiyue Mountain.

This area is adjacent to the center of hydrocarbon source, and located in the slope area of paleouplift, close to the platform margin, and developed with beach body. It is difficult to identify the traps because of the poor quality of seismic data. And the preservation condition is poor because of the complex structures.

Seven rows of high-steep structures extending from northeast to southwest in the eastern Sichuan Basin are developed, which are shown as typical ejective high-steep structures, by the structural comprehensive interpretation of seismic data and the structural maps of top Longwangmiao Formation in under-salt strata of Cambrian and top Dengying Formation in Sinian. Among them, the tectonic position of the central and western regions is relatively high, and the southern region is relatively low, while the northern region is the lowest. According to the comprehensive analysis, the first five rows of major structures developed from west to east in the central and western regions are the favorable target areas for the next exploration of under-salt strata in Sinian-Cambrian. The five rows of structural zones can be evaluated as two types of favorable exploration areas by the structural scale, burial depth, trap implementation and preservation conditions. The first three rows of structures are favorable exploration zones of Type-I, including the north Huayingshan-Honghuadian-Sihaishan structure belt in the first row, Liangshuijing-Pubaoshan-Qilixia structure belt and Wenquanjing structure belt in the second row, and Datianchi structure belt in the third row. The structural traps are developed with large scale, relatively shallow buried depth, relatively high degree of trap implementation and good preservation condition. The favorable exploration area of Type-II includes the behind two rows, the Nanmenchang structure belt in the fourth row and the Yunanchang structure belt in the fifth row. These structural traps are developed with large scale, relatively deep buried depth, relatively high degree of trap implementation and good preservation condition (Figs. 5.21 and 5.22).

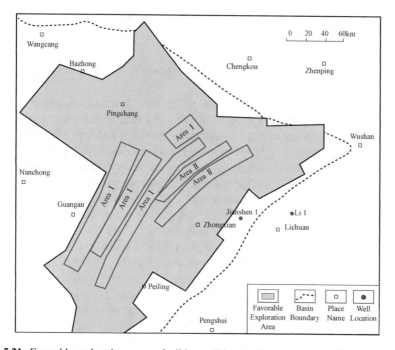

Fig. 5.21 Favorable exploration zone and wildcat well location in structure map of top Longwang-miao Formation in the East Sichuan Basin

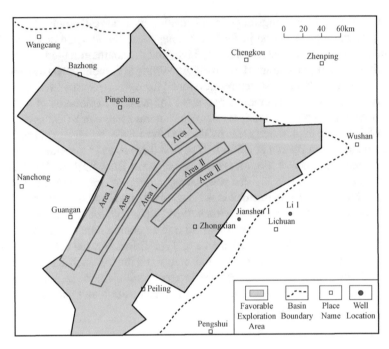

Fig. 5.22 Favorable exploration zone and wildcat well location in structure map of top Dengying Formation in the East Sichuan Basin

② Evaluation of the upper-salt Favorable Zone in the East Sichuan Basin

The reservoirs in Sinian-Lower Paleozoic of the East Sichuan Basin can be divided into two gas-bearing systems, such as upper-strata system and under-salt system, by the wide development of gypsum-salt rocks in middle and lower Cambrian. The major exploration layer in upper-salt gas-bearing system is the Xixiangchi Formation in Cambrian. The oil and gas accumulation is developed by the lateral connection of reservoirs in Xixiangchi Formation and high-quality source rocks in Silurian through the faults. The oil and gas accumulation model in upper-salt system by lateral connection can be widely developed as the Xixiangchi Formation generally thrusts over the strata of Silurian in the East Sichuan Basin (Fig. 5.23). The structural traps of upper-salt gas-bearing system are developed with lateral connection of faults in the six rows of structural belts developed from west to east in the East Sichuan Basin. Baohechang, Fuchengzhai and Tieshan structures are developed in the first row; Xiangguosi, Tongluoxia, Jiufengsi, Liangshuijing, Pubaoshan, Leiyinpu, Qilixia and Wenquanjing structures are developed in the second row; Mingyuexia and Datianchi structures are developed in the third row; Nanmenchang and Fengshengchang structures are developed in the fourth row; Yunanchang, Huangnitang, Changshou and

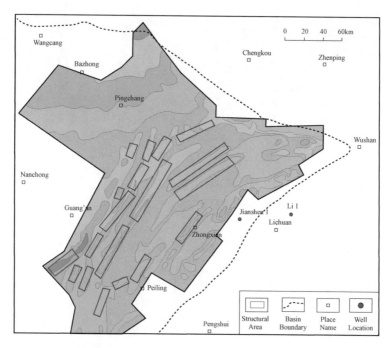

Fig. 5.23   Structure map of the top Xixiangchi Formation and the upper-salt favorable exploration area in the east Sichuan Basin

Goujiachang structures are developed in the fifth row; and Dachiganjing structure is developed in the sixth row. Through four aspects of evaluation: the Xixiangchi Formation and source rocks in Silurian are fully connected; the structural traps are preserved complete with large scale; and the fault distance is moderate, without breaking to the surface, and the preservation conditions are good; the seismic data is high quality, the structure is implemented and the burial depth is moderate. Four structures, such as Yunanchang, Datianchi, Nanmenchang and Wenquanjing, are selected as the upper-salt favorable exploration zones, by the evaluation of six rows of structures, which meet the oil and gas accumulation model of lateral connection in upper-salt system in the East Sichuan Basin.

Reference

Xunan, Huang, Li Huaqi, Wang Xizhu, et al. 2010. Evaluation indexes for completed oil and gas exploration projects and quantified evaluation [J]. *International Petroleum Economics* 18 (1): 57–61.

Plate I

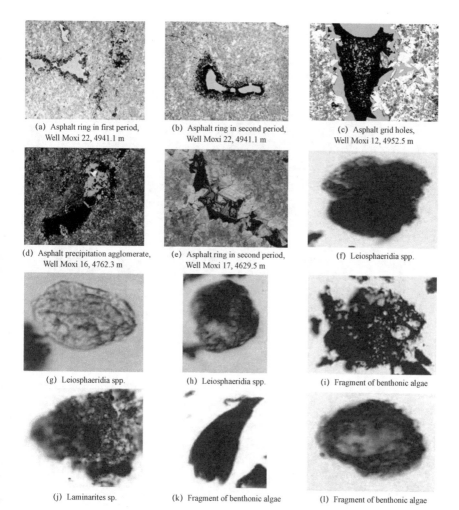

(a) Asphalt ring in first period, Well Moxi 22, 4941.1 m

(b) Asphalt ring in second period, Well Moxi 22, 4941.1 m

(c) Asphalt grid holes, Well Moxi 12, 4952.5 m

(d) Asphalt precipitation agglomerate, Well Moxi 16, 4762.3 m

(e) Asphalt ring in second period, Well Moxi 17, 4629.5 m

(f) Leiosphaeridia spp.

(g) Leiosphaeridia spp.

(h) Leiosphaeridia spp.

(i) Fragment of benthonic algae

(j) Laminarites sp.

(k) Fragment of benthonic algae

(l) Fragment of benthonic algae

© Petroleum Industry Press 2021
S. Hu and T. Wang, *Deep-Buried Large Hydrocarbon Fields Onshore China:
Formation and Distribution*, https://doi.org/10.1007/978-981-16-2285-4

Plate I

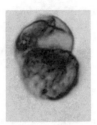

(m) Synsphaeridium sp.　　　　(n) Synsphaeridium sp.　　　　(o) Palaeolyngbya sp

Plate II

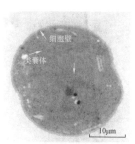

(a) Microcystis, with NaNO₃ concentration of 1.5 g/L

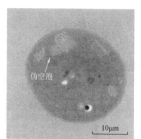

(b) Microcystis, with NaNO₃ concentration of 3.0 g/L

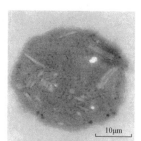

(c) Microcystis, with NaNO₃ concentration of 6.0 g/L

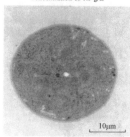

(d) Microcystis, with NaNO₃ concentration of 12 g/L

(e) Puple Chlamydomonas, with KNO₃ concentration of 0.75 g/L

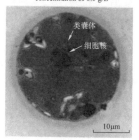

(f) Puple Chlamydomonas, with KNO₃ concentration of 1.5 g/L

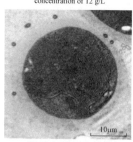

(g) Puple Chlamydomonas, with KNO₃ concentration of 3.0 g/L

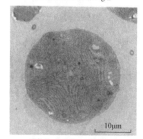

(h) Puple Chlamydomonas, with KNO₃ concentration of 6.0 g/L

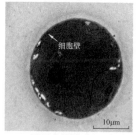

(i) Puple Chlamydomonas, with U₃O₈ concentration of 0

© Petroleum Industry Press 2021
S. Hu and T. Wang, *Deep-Buried Large Hydrocarbon Fields Onshore China: Formation and Distribution*, https://doi.org/10.1007/978-981-16-2285-4

Plate II

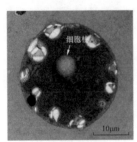

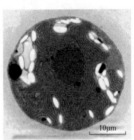

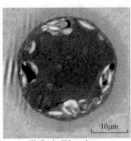

(j) Puple Chlamydomonas,
with U₃O₈ concentration of 3 mg/L

(k) Puple Chlamydomonas,
with U₃O₈ concentration of 15 mg/L

(l) Puple Chlamydomonas,
with U₃O₈ concentration of 75 mg/L

Plate III

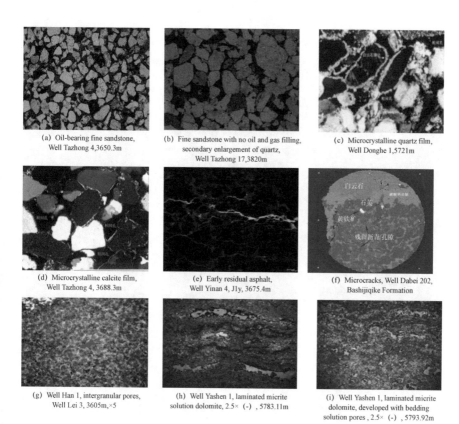

(a) Oil-bearing fine sandstone, Well Tazhong 4,3650.3m

(b) Fine sandstone with no oil and gas filling, secondary enlargement of quartz, Well Tazhong 17,3820m

(c) Microcrystalline quartz film, Well Donghe 1,5721m

(d) Microcrystalline calcite film, Well Tazhong 4, 3688.3m

(e) Early residual asphalt, Well Yinan 4, J1y, 3675.4m

(f) Microcracks, Well Dabei 202, Bashijiqike Formation

(g) Well Han 1, intergranular pores, Well Lei 3, 3605m,×5

(h) Well Yashen 1, laminated micrite solution dolomite, 2.5× (-) , 5783.11m

(i) Well Yashen 1, laminated micrite dolomite, developed with bedding solution pores , 2.5× (-) , 5793.92m

© Petroleum Industry Press 2021
S. Hu and T. Wang, *Deep-Buried Large Hydrocarbon Fields Onshore China: Formation and Distribution*, https://doi.org/10.1007/978-981-16-2285-4

Bibliography

Behar, F., S. Kressmann, J.L. Rudkiewicz, et al. 1992. Experimental simulation in a confined system and kinetic modelling of kerogen and oil cracking. *Organic Geochemistry* 19 (1–3): 173–189.

Hansheng, Qiao. 2002. *Petroleum Geology of Deep Zong in Eastern China.* Beijing: Petroleum Industry Press.

Hao, F., H.Y. Zou, Z.S. Gong, et al. 2007. Hierarchies of overpressure retardation of organic matter maturation: case studies from petroleum basins in China. *AAPG Bulletin* 91 (10): 1467–1498.

Jian, Wang, Zeng Zhaoguang, Chen Wenxi, et al. 2006. New evidence of Neoproterozoic rift sedimentary onlap and its opening age in South China. *Sedimentary Geology and Tethyan Geology* 26 (4): 1–7.

Shixin, Zhou, Wang Xianbin, Tuo Jincai, et al. 1991. New progress in deep oil and gas geochemistry. *Natural Gas Geoscience* 6: 9–15.

Xin, Shi, Dai Jinxing, and Zhao Wenzhi. 2005. Analysis of deep oil and gas reservoirs exploration prospect. *China Petroleum Exploration* 10 (1): 1–10.

Zha Ming, Qu., and Zhang Weihai Jiangxiu. 2002. Mechanism of hydrocarbon generation delay in deep ultrahigh pressure. *Petroleum Exploration and Development* 29 (1): 19–23.

Zhengwu, Guo, Deng Kangling, and Han Yonghui. 1996. *Formation and evolution of the Sichuan Basin.* Beijing: Geological Publishing House.

Zhongjian, Qiu, Kang Zhulin, and He. Wenyuan. 2002. The recent new discoveries of oil and gas fields in China and their implication. *Acta Petrolei Sinica* 23 (4): 1–6.

© Petroleum Industry Press 2021
S. Hu and T. Wang, *Deep-Buried Large Hydrocarbon Fields Onshore China: Formation and Distribution*, https://doi.org/10.1007/978-981-16-2285-4

Printed in the United States
by Baker & Taylor Publisher Services